Data Converters for Wireless Standards

THE KLUWER INTERNATIONAL SERIES
IN ENGINEERING AND COMPUTER SCIENCE

ANALOG CIRCUITS AND SIGNAL PROCESSING
Consulting Editor: **Mohammed Ismail.** *Ohio State University*

Related Titles:

AUTOMATIC CALIBRATION OF MODULATED FREQUENCY SYNTHESIZERS
D. McMahill
ISBN: 0-7923-7589-0
MODEL ENGINEERING IN MIXED-SIGNAL CIRCUIT DESIGN
S. Huss
ISBN: 0-7923-7598-X
CONTINUOUS-TIME SIGMA-DELTA MODULATION FOR A/D CONVERSION IN RADIO RECEIVERS
L. Breems, J.H. Huijsing
ISBN: 0-7923-7492-4
DIRECT DIGITAL SYNTHESIZERS: THEORY, DESIGN AND APPLICATIONS
J. Vankka, K. Halonen
ISBN: 0-7923 7366-9
SYSTEMATIC DESIGN FOR OPTIMISATION OF PIPELINED ADCs
J. Goes, J.C. Vital, J. Franca
ISBN: 0-7923-7291-3
OPERATIONAL AMPLIFIERS: Theory and Design
J. Huijsing
ISBN: 0-7923-7284-0
HIGH-PERFORMANCE HARMONIC OSCILLATORS AND BANDGAP REFERENCES
A. van Staveren, C.J.M. Verhoeven, A.H.M. van Roermund
ISBN: 0-7923-7283-2
HIGH SPEED A/D CONVERTERS: Understanding Data Converters Through SPICE
A. Moscovici
ISBN: 0-7923-7276-X
ANALOG TEST SIGNAL GENERATION USING PERIODIC ΣΔ-ENCODED DATA STREAMS
B. Dufort, G.W. Roberts
ISBN: 0-7923-7211-5
HIGH-ACCURACY CMOS SMART TEMPERATURE SENSORS
A. Bakker, J. Huijsing
ISBN: 0-7923-7217-4
DESIGN, SIMULATION AND APPLICATIONS OF INDUCTORS AND TRANSFORMERS FOR Si RF ICs
A.M. Niknejad, R.G. Meyer
ISBN: 0-7923-7986-1
SWITCHED-CURRENT SIGNAL PROCESSING AND A/D CONVERSION CIRCUITS: DESIGN AND IMPLEMENTATION
B.E. Jonsson
ISBN: 0-7923-7871-7
RESEARCH PERSPECTIVES ON DYNAMIC TRANSLINEAR AND LOG-DOMAIN CIRCUITS
W.A. Serdijn, J. Mulder
ISBN: 0-7923-7811-3
CMOS DATA CONVERTERS FOR COMMUNICATIONS
M. Gustavsson, J. Wikner, N. Tan
ISBN: 0-7923-7780-X
DESIGN AND ANALYSIS OF INTEGRATOR-BASED LOG -DOMAIN FILTER CIRCUITS
G.W. Roberts, V. W. Leung
ISBN: 0-7923-8699-X
VISION CHIPS
A. Moini
ISBN: 0-7923-8664-7
COMPACT LOW-VOLTAGE AND HIGH-SPEED CMOS, BiCMOS AND BIPOLAR

DATA CONVERTERS FOR WIRELESS STANDARDS

CHUNLEI SHI
Qualcomm CDMA Technologies
Qualcomm Inc.
San Diego, CA 92121, USA

MOHAMMED ISMAIL
Analog VLSI lab.
The Ohio State University
Columbus, OH 43210, USA

Springer Science+Business Media, LLC

ISBN 978-1-4757-7584-6 ISBN 978-0-306-48006-5 (eBook)
DOI 10.1007/978-0-306-48006-5

Library of Congress Cataloging-in-Publication Data

Printed on acid-free paper.

Contents

List of Figures ix
List of Tables xiii
Preface xv

1. INTRODUCTION 1
   1 Background 1
   2 Motivation and Goals 4
   3 Organization 5

2. OVERVIEW OF WIRELESS RECEIVER ARCHITECTURES 7
   1 Introduction 7
   2 Receiver Architecture 7
      2.1 Superheterodyne Architecture 8
      2.2 Zero-IF Architecture 9
      2.3 Low-IF Architecture 10
      2.4 Wideband IF Double Conversion Architecture 10
   3 Multi-standard Receiver Architecture 12
   4 Summary 13

3. LOW POWER ADC DESIGN 15
   1 Introduction 15
   2 Characterizations of ADC 15
   3 Review of ADC Architectures 17
      3.1 Flash ADC 17
      3.2 Interpolating and Folding ADC 17
      3.3 Two-Step ADC 19
      3.4 Oversampling ADC 19

3.5 Pipeline ADC 20

4 Overview of Pipeline ADC Designs 21

4.1 Key Building Blocks 21

4.1.1 Switched-Capacitor DAC and Residue Amplifier 22

4.1.2 Sub-ADC 22

4.1.3 Sample-and-Hold 23

4.2 Digital Error Correction 23

4.3 Design Considerations 26

4.3.1 Size of Capacitors 26

4.3.2 Capacitor Matching 27

4.3.3 Amplifier Architecture 28

4.3.4 Amplifier Requirements 28

4.3.5 Error Tolerances 31

5 Power Optimization Techniques 32

5.1 Optimizing the Stage Resolution 32

5.2 Dynamic Comparator 33

5.3 Capacitor/Amplifier Scaling 34

5.4 Dynamic Biasing 34

6 Summary 37

4. PROTOTYPE DESIGN:
ADC FOR WLAN(DSSS)/WCDMA 39

1 Introduction 39

2 Applications 39

3 Architecture 40

4 High-Speed OTA 42

4.1 OTA Requirements 42

4.2 OTA Topology 43

4.2.1 DC Gain 44

4.2.2 Gain-Bandwidth 44

4.2.3 Slew Rate 44

4.2.4 Thermal Noise 45

4.3 CMFB 45

4.4 Results 45

5 Comparator 46

6 Clock Generator 47

7 Smart-Biasing Technique 49

8 Prototype Implementation 49

9 Performance of the Prototype ADC 49
10 Summary 50

5. DESIGN CONSIDERATIONS OF LOW VOLTAGE ADCS 53
1 Introduction 53
2 Challenges in Low Voltage ADC Design 54
2.1 Low Voltage CMOS Switches 56
3 Low-V_T Devices 58
4 Clock Boosting 59
5 Switched-Opamp 60
6 A Modified Switched-opamp Technique 62
6.1 Input Stage 62
6.2 High Speed Design Techniques 66
7 Summary 68

6. ADC FOR BLUETOOTH/WLAN(FHSS)/HOMERF 71
1 Applications 71
2 System Level Design 73
3 A Switched-opamp MDAC 74
4 Opamp Design 75
4.1 DC Gain 76
4.2 Frequency Response 77
4.3 Slew Rate 78
4.4 Noise 79
4.5 Common-Mode Feedback (CMFB) 80
5 Comparator Design 81
6 Clock Generator 83
7 Digital Correction Circuit 84
8 Performance 84
9 Summary 86

7. HIGH-RESOLUTION DAC DESIGN TECHNIQUES 91
1 Review of DAC Architectures 91
1.1 Current Steering DAC 91
1.2 Switched-Capacitor DAC 93
1.3 Resistor String DAC 94
2 Intrinsic Matching of Resistor-String DAC 94
2.1 Resistor Matching Model 95

2.2 Design Techniques for Improved Resistor Matching 97
2.2.1 Reducing Random Errors 97
2.2.2 Reducing Gradient Errors 99
3 Summary 101

8. CONTROL DAC FOR 3G (UMTS) TRANSCEIVERS 103
1 Applications 103
2 Design Specifications 104
3 Architecture 104
4 Design of Resistor Strings 107
5 Class-AB Output Buffer 108
6 Deglitching Circuit 110
7 Performance of the Prototype DAC 112
8 Summary 114

9. CONCLUSION 117

Index 125

List of Figures

1.1	Die Size Change of Mixed-Signal Systems with Technology Scaling-down	3
1.2	Different Locations of ADCs in Wireless Receivers	3
2.1	Wireless Receiver Classification	8
2.2	Conventional Super-Heterodyne Receiver Architecture	9
2.3	Zero-IF Receiver Architecture	9
2.4	Low-IF Receiver Architecture	11
2.5	Wideband IF Double Conversion Receiver Architecture	11
2.6	A GSM/DECT/WCDMA Multi-standard Receiver	12
3.1	Static Performance of ADC	16
3.2	Flash ADC	18
3.3	Interpolating and Folding ADC	18
3.4	Two-Step ADC	19
3.5	Oversampling ADC	20
3.6	Block Diagram of a Pipeline ADC	21
3.7	MDAC	22
3.8	Comparator	23
3.9	Sample-and-Hold	24
3.10	Digital Error Correction	25
3.11	Maximum SNR versus Different C_s	27

3.12	Amplifier Architecture 1 and 2	29
3.13	Amplifier Architecture 3 and 4	29
3.14	Dynamic Comparator	33
3.15	Basic Idea of Dynamic Biasing	35
3.16	A Simplified Example of Dynamically Biased OTA	36
3.17	Timing Diagram of ϕ_1, ϕ_2 and V_{ctrl}	37
4.1	Block Diagram of the 1.5b/stage Pipeline ADC	41
4.2	A 1.5b Pipeline Stage	42
4.3	A Simple Telescopic OTA	43
4.4	Switched-Capacitor Common-Mode Feedback Circuit	46
4.5	Timing Diagram for Comparator	47
4.6	Clock Generator	48
4.7	Chip Photograph of the Experimental Pipeline ADC	50
4.8	Power Spectrum of the ADC Output	51
5.1	A Typical Switched-Capacitor S/H	55
5.2	DC Operating Point Requirement for the Input and Output Stage of an OPAMP	56
5.3	A Typical CMOS Switch and its Conductance	57
5.4	A Locally Bootstrapped Switch Circuit	60
5.5	A Typical Switched-capacitor Integrator	61
5.6	A Original Switched-Opamp Integrator	62
5.7	The Conceptual Scheme of the Active Series Switch	63
5.8	A Passive Input Interface with Part of the Pipeline Stage	64
5.9	The Proposed Switched-opamp S/H with Gain Stage	65
5.10	The Switchable Output Stage of Opamp	67
5.11	The Two-stage Opamp Used in SO Circuits	68
6.1	A Bluetooth Receiver	72
6.2	Block Diagram of the 8-bit Pipeline ADC	73
6.3	Block Diagram of the Switched-opamp MDAC	76
6.4	The Designed Two-stage Switchable Opamp	78
6.5	The Common-Mode Feedback Circuit	81
6.6	The Proposed Comparator	82
6.7	The Reference Voltage Generator	82

6.8 The Dynamic Latch Used in the Comparator 83
6.9 The Clock Waveform 83
6.10 The Clock Generator 84
6.11 The Layout of the Implemented Pipeline ADC 85
6.12 Simulated Frequency Response of Opamp 88
6.13 Simulated Transient Response of Opamp 88
6.14 Power Spectrum of ADC Output 89
7.1 A Typical Current-Steering DAC Based on Unary Current Cell Matrix 92
7.2 A Basic Switched-Capacitor DAC 93
7.3 A 3-bit Resistor-String DAC 95
7.4 A Single Resistor String 96
7.5 Dividing an Unit Resistor into 4 Parallel Sub-Resistors 97
7.6 Modified Unit Resistor Connection 98
7.7 Yield vs. Unit Resistor Standard Deviation 99
7.8 Linear Error Compensated Layout 100
7.9 Modified Layout Arrangement 101
7.10 INL of Different Layout Schemes 102
8.1 A 4-bit Folded Resistor String DAC 105
8.2 Multiple Resistor String DAC (4bit) 106
8.3 A Modified Multiple Resistor String DAC (4bit) 107
8.4 A Multiple Resistor String DAC with Two Compensation Currents 108
8.5 The DAC Nonlinearity Caused by the Load Effect of the LSB String 109
8.6 The Output Voltage Buffer Used in the DAC 110
8.7 Open Loop Gain and Phase of the Output Voltage Buffer 111
8.8 The Proposed Deglitching Circuit 112
8.9 Simulated Nonlinearity of the DAC 113
8.10 The Major Glitch at the DAC Output 114

List of Tables

1.1 Evolution of CMOS digital processing for embedded DSP and RISC cores. [Source: National Technology Roadmap for Semiconductors, Semiconductor Industry Association, USA, Jan. 1998.] 2
4.1 Summary of WLAN (DSSS) and WCDMA Standards 40
4.2 Performance Summary of the Pipeline ADC 50
5.1 SIA Technology Roadmap 1999 54
6.1 Summary of Bluetooth, WLAN (FHSS), and HomeRF Standards 71
6.2 Required Specifications of the First Opamp 74
6.3 Performance Summary of the Switched-opamp Pipeline ADC 86
8.1 Specifications of the Prototype DAC 104
8.2 Performance Summary of the Resistor String DAC 115

Preface

Wireless communication is witnessing tremendous growth with proliferation of different standards covering wide, local and personal area networks (WAN, LAN and PAN). The trends call for designs that allow 1) smooth migration to future generations of wireless standards with higher data rates for multimedia applications, 2) convergence of wireless services allowing access to different standards from the same wireless device, 3) inter-continental roaming. This requires designs that work across multiple wireless standards, can easily be reused, achieve maximum hardware share at a minimum power consumption levels particularly for mobile battery-operated devices.

All this calls for higher levels of system integration of both the radio and the digital baseband parts. It also calls for radio design solutions with mixed signal strategies that take full advantage of technology scale-down by moving functions, such as channel select filtering, modulation and demodulation, to the digital domain. Central to achieving these goals is the design of data converters for these emerging standards in the context of technology and market trends.

The book presents the design of such converters and introduces the underlying circuit design principles. As such the book will serve as a reference for IC and mixed signal designers, design managers and project leaders in industry, particularly those in the wireless semiconductor industry. The book could also serve as a reference or a text for a first year graduate course on the subject for electrical and/or computer engineering majors.

After a brief introduction of radio transceiver architectures in Chapter 2, the book deals with power optimized analog to digital converters (ADCs) in Chapter 3 where several power optimization techniques used in pipeline ADCs are analyzed and a novel technique based on dynamic biasing is introduced. An experimental prototype ADC for WLAN and WCDMA standards, based on the theory in Chapter 3, is developed in Chapter 4 in a 0.5 micron CMOS technology.

Chapter 5 deals with low voltage ADC design where a modified switched-OpAmp technique suitable for System-on-Chip (SoC) design is proposed. The technique increases the maximum sampling frequency and utilizes a novel design for the input stage. Based on this material, Chapter 6 presents a complete design of a 1.5V 8-bit 20MS/s pipeline ADC in 0.18 micron technology where the target applications are Bluetooth, HomeRF and WLAN standards based on frequency hopping spread spectrum coding.

In Chapter 7, a technique to improve the intrinsic matching of resistor-string digital-to-analog converters (DACs) is described. The technique does not use trimming or calibration. This technique is demonstrated in Chapter 8 in the design of a high resolution control DAC for 3G UMTS transceivers. Chapter 9 provides some concluding remarks.

This book has its roots in the doctoral dissertation work of the first author at the Analog VLSI Lab, The Ohio State University. We would like to thank all those who supported us at the Analog VLSI Lab and at other locations including the Radio Electronics Lab at the Swedish Royal Institute of Technology, The 3G Wireless Group at Intel Corporation, Phoenix and Spirea AB, Stockholm.

Chunlei Shi
Mohammed Ismail

Columbus, Ohio
September 2001

This book is dedicated to
Xinyan & Lincai Shi, Xinyi
and
Sameha, Ismail, Sr., Tuula,
Ismail, Jr. and Omar

Chapter 1

INTRODUCTION

1. Background

The wireless communication market has been experiencing tremendous growth and will continue to do so in the next decade. It is expected that over 450 million cellular phones will be sold worldwide in 2001, and the number of service users will increase to more than 1.8 billions by 2010 [Chaudhury et al., 1999]. Other applications including wireless local area network (WLAN) and Bluetooth will also have huge markets in the near future. For Bluetooth only, conservative estimates foresee several hundred million devices in the next five years [Haartsen and Mattisson, 2000].

In the design of mobile wireless communication systems, power consumption and form factor are two major engineering concerns [Abidi et al., 2000]. The highly competitive market demands low cost, low power and small form factor devices.

Mobile systems are usually powered by a battery which is often the largest and most expensive component in the device [Baltus and Dekker, 2000]. To prolong the battery lifetime while keeping the device at reasonable size, mobile devices must use power-conscious designs, which usually require power optimization at all levels of the system hierarchy. At higher levels, communication protocols should be optimized for low power consumption; while at lower levels, all circuit architecture and building blocks should adopt power conscious designs, including the optimization of signal processing functions and their implementation in the analog or digital domain.

Data converters including analog-to-digital converters (ADCs) and digital-to-analog converters (DACs) are essential building blocks in almost all mixed-signal systems. They provide an interface between the analog world and the digital world. Over the past decades, digital signal processing (DSP) has been proved to be a cost-effective method over its analog counterpart, and therefore it is desirable to push more signal processing into the digital domain. As a consequence, demanding requirements are put on data converter designs.

Shown in Table 1.1 is the evolution roadmap of CMOS digital technology from the Semiconductor Industry of Association (SIA). As can be seen, digital circuits get direct benefits in terms of device density and computation capability with the scaling-down of technology (known as Moore's law). On the other hand, analog circuits do not benefit from scaling. For example, since the vertical plate spacing does not decrease as transistor dimension shrinks in CMOS technologies, a given value of a parallel plate capacitor tends to consume a fixed area. Therefore an analog filter when designed in a 0.5μm technology will use the same device sizes and areas as in a 0.25μm technology. But if it is moved to the digital domain and implemented as a digital filter, then it will ride the scale down curve (Figure 1.1).

Table 1.1. Evolution of CMOS digital processing for embedded DSP and RISC cores. [Source: National Technology Roadmap for Semiconductors, Semiconductor Industry Association, USA, Jan. 1998.]

Year	1995	1997	1999	2001	2003	2006	2012
Gate length (μm)	0.35	0.25	0.18	0.15	0.13	0.10	0.07
Wafer size (inch)	8	8	8	12	12	12	12
Supply voltage (V)	3.3	1.8-2.5	1.5-1.8	1.2-1.5	1.2-1.5	0.9-1.2	0.5-0.8
Clock frequency (MHz)	100	200	400	600	700	900	1500
DSP MIPS	N/A	100	200	400	900	1000	N/A
Typical die size (mm^2)	250	300	300	360	430	520	630

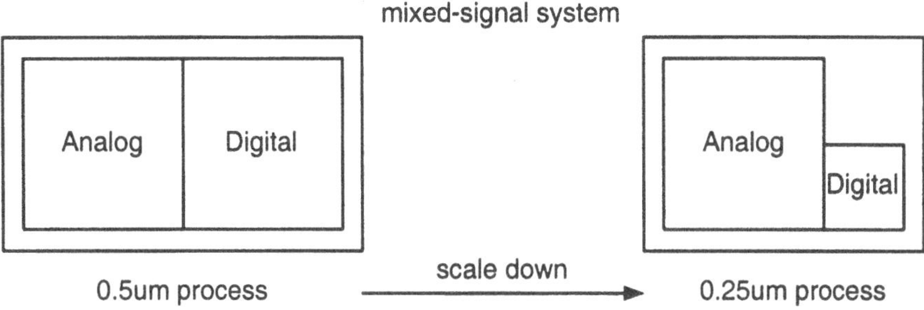

Figure 1.1. Die Size Change of Mixed-Signal Systems with Technology Scaling-down

In wireless communication systems, complex filtering and decoding schemes can be implemented more easily using DSP, and if certain functions of an analog baseband chain (such as channel select filtering, dc offset compensation, variable gain amplification, etc) are moved to the digital domain, then the total solution will benefit from the technology scaling-down and the cost will be greatly reduced (Figure 1.2). This solution will call for very good data converter designs since it is very challenging for both the dynamic range, speed and power consumption, supply voltage, etc.

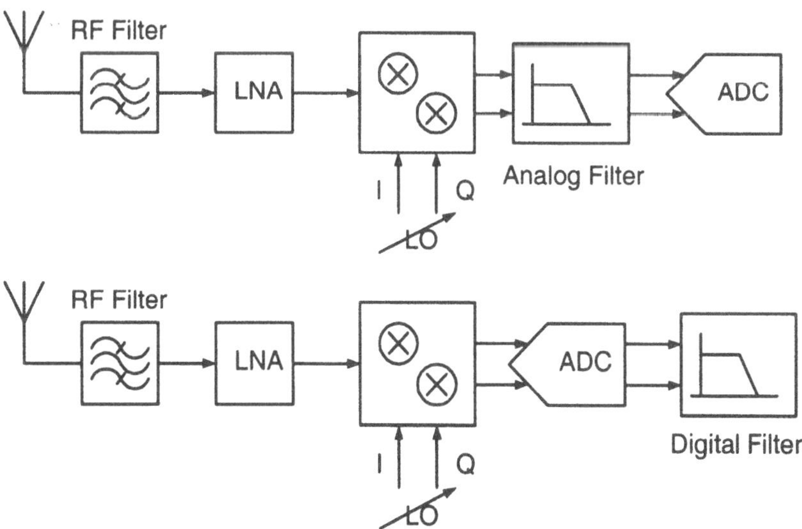

Figure 1.2. Different Locations of ADCs in Wireless Receivers

The expanding growth of wireless communications has also led to the proliferation of different standards. Although Global System for Mobile communication (GSM) from Europe is dominating the second generation (2G) mobile systems, there are several other incompatible standards worldwide including IS-54, IS-136 and IS-95 from North America and PDC from Japan. These different standards employ different multiple access techniques, different modulation and coding schemes, and different channel bandwidth assignments, thus prevent global roaming. The third generation (3G) wireless systems aiming to integrate multiple applications and provide wider compatibility will also be implemented within the next few years. It is clear that multi-standard cellular networks will co-exist for a long time to serve the wireless market, which calls for the need of developing a single transceiver capable of operating at different modes with multi-standard support features.

Different wireless standards have different channel bandwidth and dynamic range requirements. For example, in GSM the filter bandwidth is only 100kHz but the dynamic range is over 80dB; while in WCDMA the filter bandwidth is up to 5MHz but the dynamic range is only around 36dB. If channel select filtering is moved to the digital domain as discussed above, then a single wideband data converter with high dynamic range is required.

High-speed high-resolution data converters usually consume large power, which becomes a major problem when they are integrated with other blocks in RF transceivers, especially in battery powered handsets.

High speed Nyquist-rate ADCs with reasonable power consumption have mainly been implemented using BiCMOS technology, and are only recently becoming available in CMOS [Brandt and Lutsky, 1999, Mehr and Singer, 1999].

2. Motivation and Goals

Clearly, power optimized high-speed high-resolution data converters will greatly reduce the total cost, power consumption and form factor of mobile devices. The benefits are two fold: 1). Such a data converter is of great help when choosing a power conscious architecture; 2). It also reduces the power consumption at the circuit level.

Besides low power consumption, low voltage operation is another important design constraint in these portable systems. This trend has been driven by several factors such as the continuous scaling-down of CMOS technologies and system-on-chip (SoC) design requirements. In order to be compatible with low-voltage systems, a new type of data converters that can operate with a single battery (1.5V) are highly desired. It is worthy to note that in analog circuits, low supply voltages do not lead to low power consumptions as in the case of their digital counterparts. On the contrary, larger power consumption may be required in order to maintain the same dynamic range with low supply voltages.

The objective of this work is to investigate new circuit techniques to design low-power/low-voltage CMOS data converters for emerging wireless standards such as WCDMA, WLAN, Bluetooth, HomeRF, etc.

The main goals can be summarized as follows:

- Explore power optimization techniques of high-speed CMOS pipeline A/D converters. Design a power optimized ADC for WLAN(DSSS)/ WCDMA direct conversion receivers.
- Investigate design techniques of low voltage analog building blocks. Design a low voltage video-rate switched-opamp pipeline ADC for Bluetooth, WLAN(FHSS), and HomeRF applications.
- Study design techniques of high-resolution DAC in standard CMOS technologies. Design a power-efficient high-resolution DAC for 3G (UMTS) transceivers.

3. Organization

This book is divided into 9 chapters.

In Chapter 2, several radio architectures are briefly reviewed.

Chapter 3 discusses the power optimization techniques of high-speed ADCs. A prototype high-speed low-power ADC design is presented in Chapter 4. The design is suitable for direct sequence spread spectrum (DSSS) WLAN standards and third generation (3G) wideband CDMA (WCDMA) standards.

In Chapter 5, design techniques of low voltage analog building blocks are studied. A prototype low voltage ADC design is presented in detail in Chapter 6. The target applications are Bluetooth, HomeRF and WLAN standards based on frequency hopping spread spectrum coding.

Chapter 7 discusses the design techniques of high resolution D/A converters. In Chapter 8, a prototype DAC design is described. The design can be used for 3G UMTS transceivers.

Finally, Chapter 9 concludes this book.

Chapter 2

OVERVIEW OF WIRELESS RECEIVER ARCHITECTURES

1. Introduction

The recent evolution in wireless communications requires the development of a low cost single chip CMOS transceiver capable of operating at multiple RF standards.

In this chapter, several radio receiver architectures are briefly reviewed, with the emphasis on the issues related to monolithic integration and multi-standard operation.

2. Receiver Architecture

Depending on the location of the ADC in the receiver chain, the basic receiver architectures can be classified accordingly into three categories as shown in Figure 2.1[Elwan et al., 2001]. In the first category, the ADC is placed in the RF section, and the RF signal at the antenna is digitized before being processed in the digital domain. This ideal "software radio" architecture is suitable for multi-standard operation by its nature. Unfortunately, this architecture is not practical even in the foreseen future semiconductor technologies, since it requires ADCs with extremely high linearity over the RF sampling frequency range (GHz).

In the second category, the ADC is put in the IF section, preceded by low noise amplifier (LNA) and mixer. Compared to the previous implementation, the ADC requirement is relaxed and is achievable with current silicon technologies. The problem is that analog filtering and image rejection requirements on the RF analog section are more difficult to achieve [Elwan et al., 2001].

The third category of receivers digitizes the signal at baseband. By implementing more analog signal processing before the ADC, these architectures relax the design requirements of the ADC, and are therefore more practical. In the next section, several architectures in this category including superheterodyne, zero-IF, low-IF, and wideband IF double conversion will be compared.

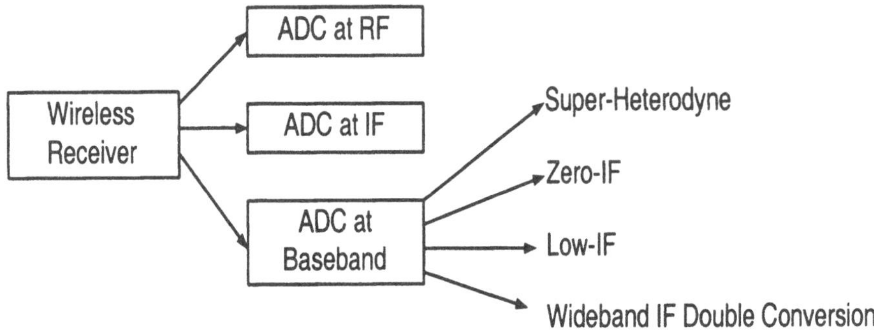

Figure 2.1. Wireless Receiver Classification

2.1 Superheterodyne Architecture

Shown in Figure 2.2 is the conventional super-heterodyne receiver architecture, which is currently used in most commercial wireless receivers. In this architecture, the RF spectrum first passes through a discrete RF filter that eliminates out-of-band energy and slightly rejects the image-band signals. The LNA amplifies the signal before another image rejection (IR) filter further attenuates the undesired signals presented at the image frequencies.

After the IR filter, a RF variable frequency synthesizer (LO1) tunes the desired band to a fixed intermediate frequency (IF) where a discrete high-Q IF filter is used to attenuate alternate channel energies. Then the signal is downconverted to the baseband by mixing with a fixed frequency synthesizer (LO2).

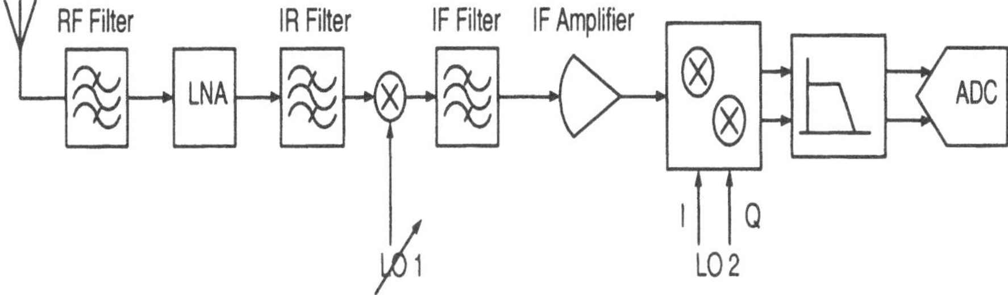

Figure 2.2. Conventional Super-Heterodyne Receiver Architecture

Although a superheterodyne receiver achieves superior performance in terms of selectivity and sensitivity, it requires high-Q high-performance off-chip discrete components to realize the IF stage. Therefore this architecture is not amenable to the highly-integrated solution required by modern potable communication systems.

2.2 Zero-IF Architecture

Transceiver architectures suitable for higher levels of integration are the zero-IF, the low-IF, and the wideband IF double conversion configurations [Gray and Meyer, 1995]. In the zero-IF (also called direct conversion or homodyne) architecture as shown in Figure 2.3, the RF spectrum is directly downconverted to the baseband, thus eliminating both the IR and IF filters. The subsequent filtering and gain amplification are all to be done in baseband at low frequencies.

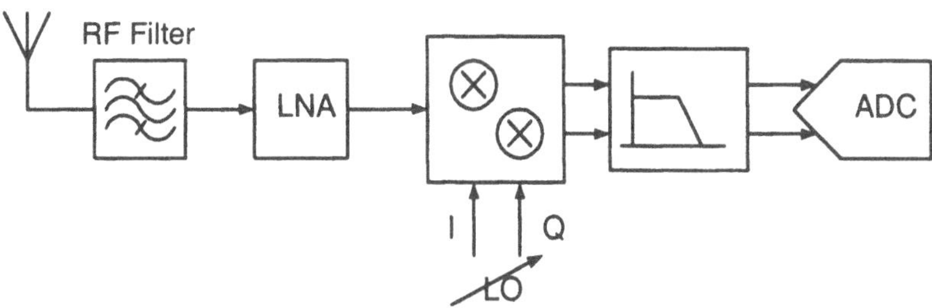

Figure 2.3. Zero-IF Receiver Architecture

A Zero-IF architecture is suitable for monolithic integration and amenable to multi-standard operation, since it requires no image rejection filter (because the IF is zero) and only a minimum RF section comprising a LNA and two mixers. Furthermore, the LNA need not drive a 50Ω load because no image rejection filter is required.

However, there exists several well-known design challenges. The first one is the DC offset problem. One important error source is the device-mismatch-induced DC offset and $1/f$ noise of the signal path itself [Gray and Meyer, 1995]. Another important source of DC offset is the LO-to-antenna leakage. Since the LO is at the same frequency as the RF carrier, it may transmit to the antenna and get reflected, resulting in self-mixing. The resulting time-varying DC offset reduces the receiver dynamic range and may even saturate subsequent baseband circuits. It should thus be removed either by analog feedback circuitry or adaptive digital cancellation. The second challenge is that this architecture requires a low phase noise frequency synthesizer which is difficult to implement with low-Q on-chip oscillators.

2.3 Low-IF Architecture

In low-IF receivers, the RF signal is translated to a low, but non-zero, frequency. Thus the DC offset and $1/f$ noise problems are eliminated. It can be regarded as a special kind of superheterodyne receiver, which converts the received RF signal to a low intermediate frequency whereby the on-chip digital or analog bandpass filtering can be used to perform channel selection. However, it reintroduces the image rejection problem. Because of the relatively low close-in image frequency, the image rejection is very difficult. Image rejection mixer techniques are required instead of image rejection filter, where the phase shift accuracy and path matching accuracy within the mixer must be extremely precise.

Shown in Figure 2.4 is the block diagram of a typical low-IF receiver [Tadjpour et al., 2001].

2.4 Wideband IF Double Conversion Architecture

An alternative architecture suitable for integration of the entire receiver is wide-band IF with double conversion receiver (WBIFDC). As

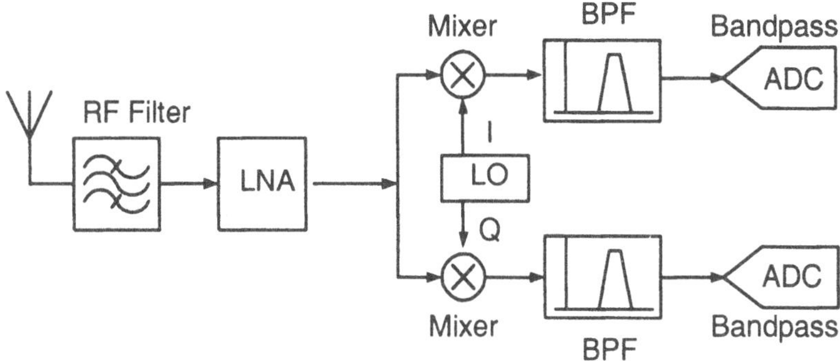

Figure 2.4. Low-IF Receiver Architecture

shown in Figure 2.5, this receiver translates the desired RF band to IF using a mixer with a fixed frequency LO. This LO is at higher frequency and can be implemented with a low phase noise and wide phase locked loop bandwidth using low-Q on-chip components. Upconverted frequency signals from the mixer are eliminated using a simple IF low-pass filter. The second mixer downconverts the IF signal to baseband using a tunable frequency synthesizer. In this architecture, since there is no local oscillator operating at the same frequency as the RF signal, the LO leakage problem that plagues direct conversion receivers and results in time-varying dc offsets is eliminated. Moreover, the LO leakage caused by re-radiation of the second local oscillator signal is not a serious problem in this architecture since it only causes a relatively constant DC offset, which may be easily removed by simple circuits [Rudell et al., 1997].

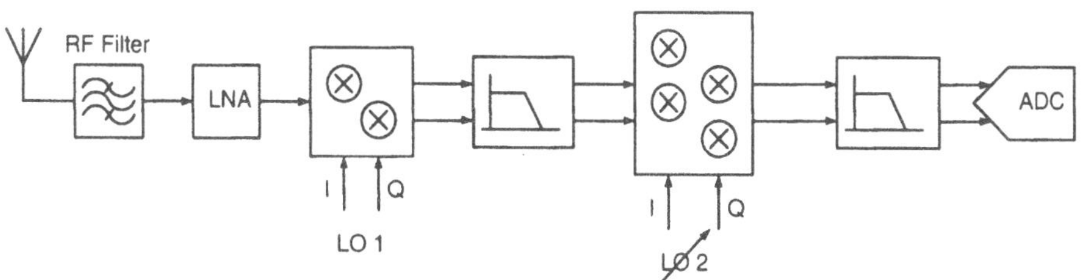

Figure 2.5. Wideband IF Double Conversion Receiver Architecture

This approach is highly desirable for monolithic integration. However, it requires the use of six high performance mixers to perform down-conversion and image-rejection function, which increases the DC power consumption considerably. The large difference among the baseband bandwidthes of multiple standards also makes the second local oscillator difficult to design.

3. Multi-standard Receiver Architecture

Among the receiver architectures discussed above, zero-IF or direct-conversion receiver is a good candidate for monolithic integration and multi-standard operation. Historically, the design challenges described in Section 2.2 made this architecture impractical for high performance wireless receivers. However, after extensive research, novel design techniques have recently been proposed to overcome those drawbacks and the direct conversion receiver has shown its potential for various RF systems.

A GSM/DECT/WCDMA multi-standard zero-IF receiver architecture is depicted in Figure 2.6 [Li, 2001]. In this architecture, three sets of band filters and LNAs are required for the band selection and low noise amplification, while the baseband filter, VGA and A/D converter are shared by the three standards.

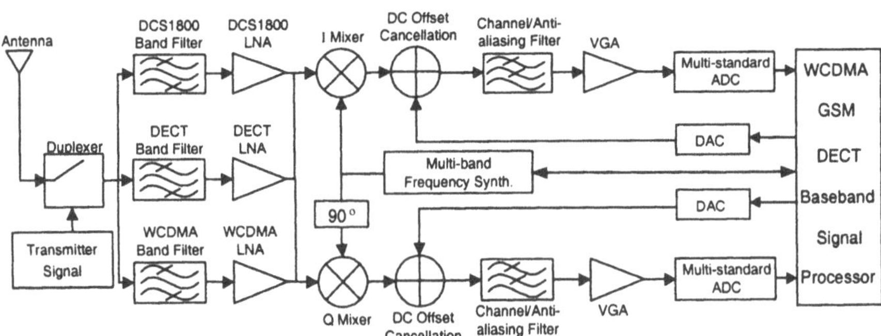

Figure 2.6. A GSM/DECT/WCDMA Multi-standard Receiver

4. Summary

In this chapter, different wireless receiver architectures are briefly introduced and compared. In the following chapters, novel techniques will be investigated to design low-power/low-voltage CMOS data converters (analog-to-digital converters and digital-to-analog converters) for several different wireless applications such as WCDMA, WLAN, Bluetooth, HomeRF, etc. It is worthy to note that although all the circuits discussed in this book are originally designed for zero-IF receivers, they can also be used in other receiver architectures.

Chapter 3

LOW POWER ADC DESIGN

1. Introduction

In this chapter, several widely used ADC architectures are first introduced and compared, among which pipeline ADC is most suitable for wideband wireless applications. Next, major design considerations of pipeline ADCs are described. Lastly, several low-power design techniques for pipeline ADCs are discussed and a novel technique is proposed.

2. Characterizations of ADC

Traditionally an ADC is characterized using its static (DC) performance. Typical parameters include resolution, offset, gain error, differential nonlinearity (DNL), and integral nonlinearity (INL), as depicted in Figure 3.1 [Jones and Martin, 1996].

1 Resolution: Defined as the number of distinct analog levels corresponding to the different digital words.

2 Offset: Defined as the deviation of $V_{0...01}$ (analog signal value corresponding to digital code $0\ldots01$) from $1/2$ LSB.

3 Gain error: Defined as the difference at the full-scale value between the ideal and actual transfer curves when the offset error has been reduced to zero.

4 INL: Measured as the maximum deviation from a straight-line approximation after the offset and gain error have been removed.

5 DNL: Defined as the maximum deviation in analog step size from 1 LSB.

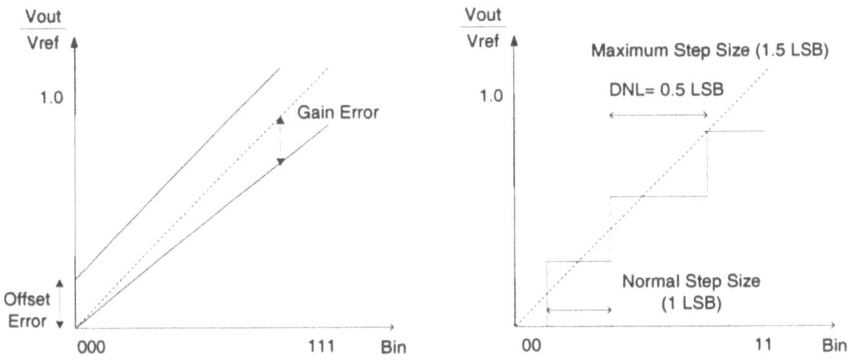

Figure 3.1. Static Performance of ADC

However, the traditional performance parameters can not be directly applied to telecommunication applications. For example, the resolution of an ADC only refers to the number of digital output bits, instead of an indication of the accuracy. Communication systems are usually characterized in the frequency domain, therefore frequency domain (AC linearity) performance of ADC should be considered [Hendriks, 1997]. These parameters characterize ADC more accurately, since all the dynamic and static nonlinearities will show themselves as distortion and noise in the frequency domain. Typical AC performances of an ADC include SFDR, SNR, SNDR, THD, etc.

1 SFDR (Spurious Free Dynamic Range): SFDR is the difference (in decibels) between the root-mean-square (rms) power of the fundamental and the largest harmonically or nonharmonically related spurious signal within a specified frequency band. This measure of dynamic range is useful since it indicates the dynamic range that can be obtained before distortions become dominant over noise.

2 SNR (Signal to Noise Ratio): SNR is the ratio of signal power to total noise power in the output of ADC within a specified frequency band.

3 SNDR (Signal to Noise plus Distortion Ratio): SNDR is the ratio of the signal power to the total noise plus distortion power in the output of ADC within a specified frequency band. This is the most important parameter when specifying a communication ADC, since it measures the degradation of a signal due to the combined effect of noise, quantization error and harmonic distortions.

4 THD (Total Harmonic Distortion): THD is the ratio of the rms sum of the total harmonics to the rms value of the fundamental at the output of an ADC. In practice, the first five or six harmonics are usually measured to calculate THD.

3. Review of ADC Architectures

In this section, several ADC architectures are briefly reviewed. Pipeline ADC is an idea choice for broadband communication systems due to its low power consumption, high speed and relatively high resolution.

3.1 Flash ADC

The flash architecture is a popular approach for designing very high-speed low-to-medium resolution (6 to 8 bits) converters [Peetz et al., 1986, Yoshii and *et al.*, 1987, Hotta and *et al.*, 1987, Gendai et al., 1991]. As shown in Figure 3.2, the input signal is fed to $(2^N - 1)$ comparators connected in parallel and is compared to a set of reference voltages generated from a resistor string. The $(2^N - 1)$ bit output code (called thermometer code) is then transformed to the N bit binary code.

Flash ADC is very fast due to its simple architecture. But it has several serious drawbacks. The number of comparators in a flash ADC, hence the chip area and power consumption, increases exponentially with the resolution N. The large number of comparators connected to the input signal also results in a large parasitic load. Furthermore, the accuracy requirement of the comparators also increases exponentially with N, requiring more complex comparator designs.

3.2 Interpolating and Folding ADC

Interpolating ADC makes use of input amplifiers to reduce input capacitance, as shown in Figure 3.3(a). This approach is often combined

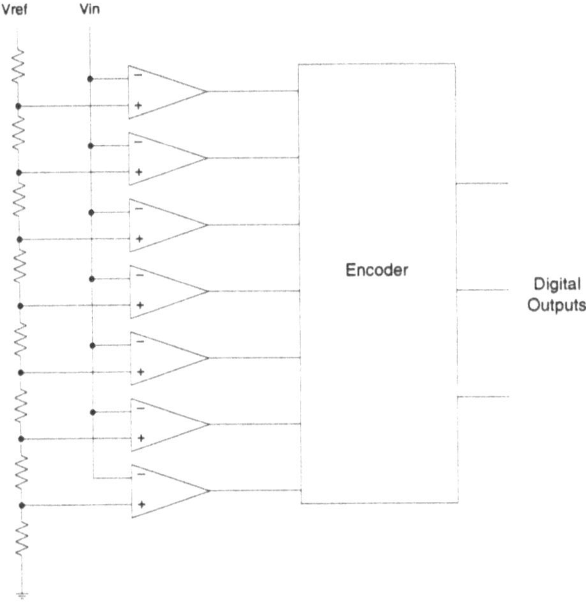

Figure 3.2. Flash ADC

with folding architecture which is illustrated in Figure 3.3(b). With the interpolating and folding architecture, the number of comparators in the ADC can be significantly reduced.

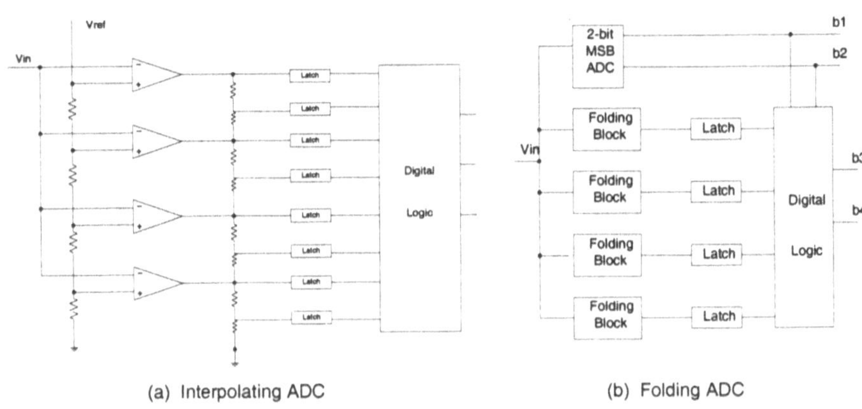

Figure 3.3. Interpolating and Folding ADC

Interpolating and folding ADCs overcome most of flash ADCs' drawbacks and have been widely used [Vorenkamp and Roovers, 1993], but this architecture is more appropriate for bipolar technologies. It does not excel regarding dynamic performance [Mehr and Singer, 1999] in CMOS technologies due to the lack of an input sample-and-hold stage.

3.3 Two-Step ADC

Figure 3.4 shows the block diagram of a two-step ADC. The MSB sub-ADC obtains $N/2$ most-significant-bits (MSBs) from the input sampled signal, and the sub-DAC block converts the MSBs back to an analog signal, while the residue between the sampled signal and this analog signal is passed to the LSB sub-ADC where the least-significant-bits (LSBs) are obtained. Compared to the flash architecture, a two-step ADC requires less comparators ($2 \times 2^{N/2}$ compared to 2^N), has less input capacitance, with relaxed requirements on the comparator. The main drawback is that it requires precise inter-stage processing, which is not an easy task at high sampling frequencies.

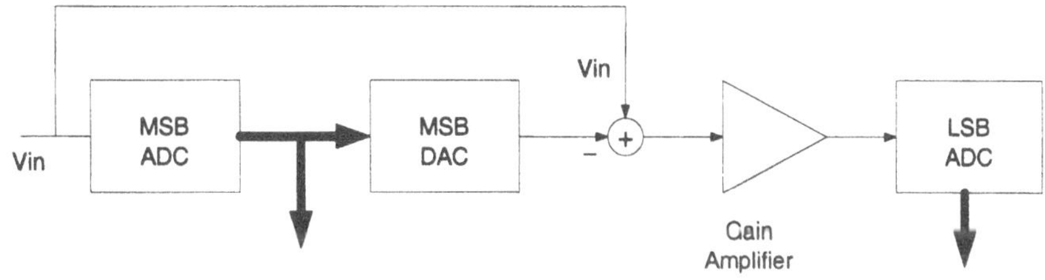

Figure 3.4. Two-Step ADC

3.4 Oversampling ADC

Recently oversampling ADCs have been widely used in high-resolution low-to-medium speed applications. This architecture has several advantages: first, it relaxes the requirements on analog circuits at the cost of more complex digital signal processing; second, it relaxes the requirement on the anti-aliasing filter before the ADC [Jones and Martin, 1996]. Shown in Figure 3.5 is a simplified block diagram of an oversampling ADC based on a first-order sigma-delta modulator. The principle of using oversampling is that by sampling the signal many times, errors due to noise and coarse quantization are averaged out. Through the use of loop-filter and feedback, the noise is shaped to high frequencies, which can be easily removed by digital filters.

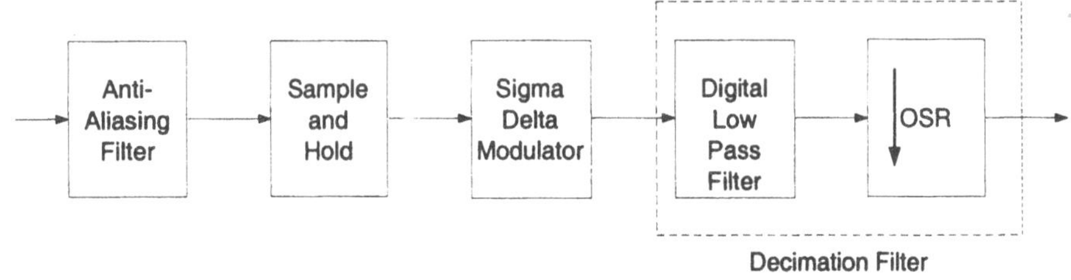

Figure 3.5. Oversampling ADC

Oversampling ADCs are traditionally limited to low-speed applications such as audio or narrow-band communication systems, since the "speed-trade-for-resolution" rule requires them to use high oversampling ratios in order to get high resolutions. Although cascode sigma-delta modulators with multi-bit DACs have the potential of achieving high resolutions with low oversampling ratios, no results have been reported for input frequencies higher than 5MHz.

3.5 Pipeline ADC

Pipeline ADC is a generalized two-step ADC with improved throughput and tolerance to comparator errors. It adds a sample-and-hold and an amplifier to each stage, which allows each stage to be immediately used to process its next input sample before the succeeding stages have finished. Figure 3.6 shows a typical pipeline ADC architecture. It consists of M identical stages, each stage samples the output from the previous stage and quantizes to k bits digital code, which is then converted back to an analog signal by the k-bit DAC. The difference between the sample signal and the restored analog signal is amplified by a gain of $G = 2^k$, then passed to the next stage. The $(M \times k)$-bit output codes are combined into the final codes in registers.

Due to the pipelining architecture, the throughput rate of an ADC is independent of the number of stages used, and the hardware cost is approximately linear with the resolution. Another advantage of the

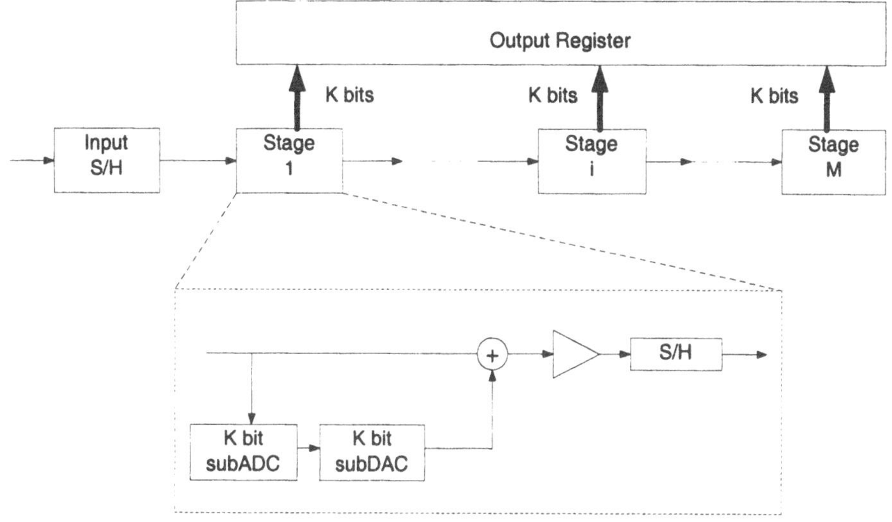

Figure 3.6. Block Diagram of a Pipeline ADC

pipeline architecture is that the accuracy requirements for the later stages are greatly relaxed compared to the first few stages, making it possible to reduce the total power consumption. Furthermore, digital correction techniques can be used to significantly reduce the sensitivity of the architecture to analog component non-idealities [Lewis and Gray, 1987].

Based on above reasons, pipeline ADC is a suitable choice for low-power high-speed high-resolution applications.

4. Overview of Pipeline ADC Designs

In the previous section, several ADC architectures are briefly reviewed, among which pipeline ADC is the most suitable candidate for low-power high-speed high-resolution applications. This section focuses on several design aspects in pipeline ADCs. Several key building blocks are first described, then sources of error are analyzed. Finally, practical design considerations are discussed.

4.1 Key Building Blocks

A detailed block diagram of a typical pipeline ADC is shown in Figure 3.6. Each stage consists of a k-bit sub-ADC, a k-bit sub-DAC, a subtraction circuit, and a residue amplifier. An input sample-and-hold is usually added in front of the first pipeline stage.

4.1.1 Switched-Capacitor DAC and Residue Amplifier

Shown in Figure 3.7 is an efficient way to implement a MDAC (multiplying DAC), which combines a sub-DAC, a subtraction circuit, and a residue amplifier in one block. It can sample and hold the input signal, generate the residue (difference between the input signal and the DAC output), and amplify the residue. The circuit works as follows: During sampling phase φ_1, the input signal is sampled onto the capacitors C_1 and C_f ; during holding phase φ_2, the capacitor C_f is switched to the amplifier output, while C_1 is switched to one of three voltages: $+V_{ref}$, $-V_{ref}$ or ground, depending on the digital code of the sub-ADC. The output voltage can be derived as:

$$V_0 = \frac{C_1 + C_f}{C_f} * V_I - \frac{C_1}{C_f} * V_{DAC} \tag{3.1}$$

where $V_{DAC} = +V_{ref}, -V_{ref}$ or 0.

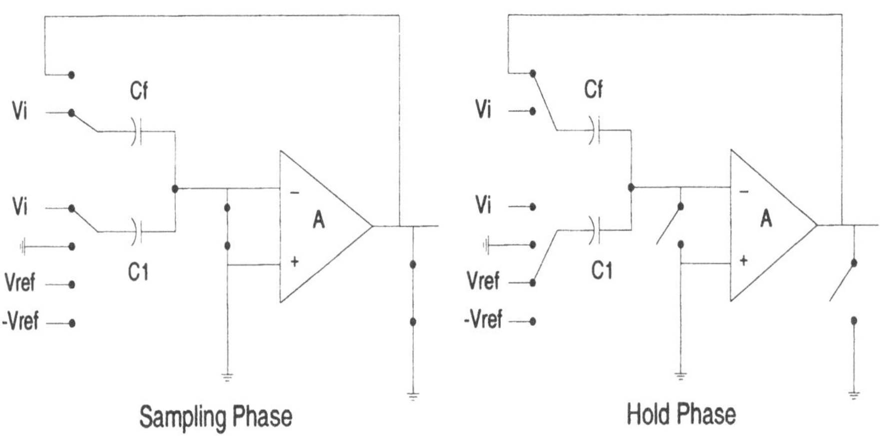

Figure 3.7. MDAC

4.1.2 Sub-ADC

The sub-ADC usually employs flash architecture, as shown in Figure 3.2. The resistor string generates the reference voltages for comparators. Depending on the accuracy requirement, a comparator usually consists of one or several preamp stages and a regenerative latch (see Figure 3.8). However, a simple dynamic latch may also be used as a comparator if digital error correction technique is employed, as will be explained in the next section.

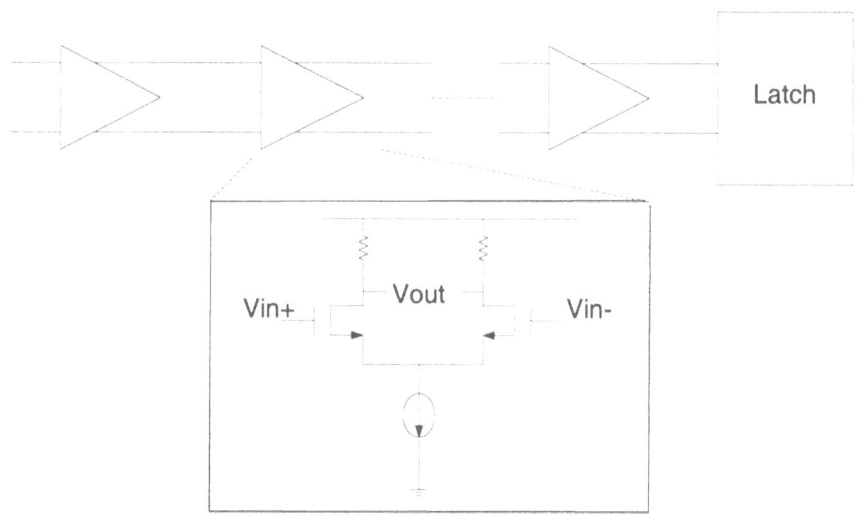

Figure 3.8. Comparator

4.1.3 Sample-and-Hold

In order to capture high-frequency input signals, most pipeline ADCs make use of a front-end sample-and-hold circuit. Figure 3.9 presents such a circuit [Shu et al., 1996]. Typically, the noise and distortion performance of the sample-and-hold limits the dynamic performance of the entire pipeline ADC.

4.2 Digital Error Correction

The primary sources of errors in a pipeline ADC are noise, offset errors (sample-and-hold, amplifier, comparator), gain errors (sample-and-hold,

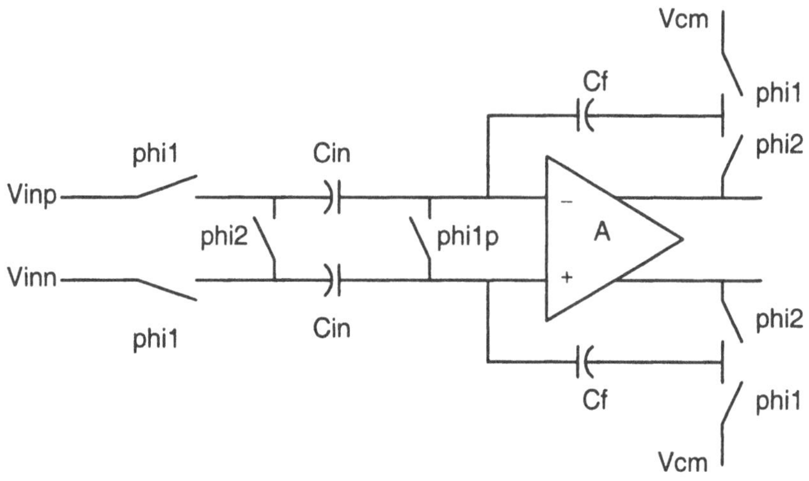

Figure 3.9. Sample-and-Hold

amplifier), sub-ADC and sub-DAC nonlinearities, and amplifier settling errors.

Thermal noise is the most fundamental source of errors in a pipeline ADC, most of which comes from sampling switches and the sample-and-hold amplifier. Sampling switch noise is also known as kT/C noise since the noise power is proportional to kT/C (k: Bolzmann constant; T: temperature; C: sampling capacitor). Noise from the sample-and-hold amplifier is also inversely proportional to a capacitor. Thermal noise is random, and the only way to reduce it in pipeline ADCs is to increase the capacitor size.

Gain errors in the sample-and-hold and the amplifier are caused by the finite amplifier gain and capacitor mismatches. Gain errors will cause nonlinearities in pipeline ADCs. DNL caused by the gain error of the first stage in a pipeline ADC is the largest and is given by [Cline, 1996]:

$$DNL(LSB) = \left(\frac{G_{ideal} - G}{G_{ideal}}\right)\left(\frac{G_{ideal} - 1}{n_c}\right)\frac{2^N}{G_{ideal}} \tag{3.2}$$

where G_{ideal} and G are the ideal and the real gain, respectively, n_c is the number of comparators per stage, and N is the ADC resolution.

Amplifier settling errors are due to the incomplete settling of amplifiers when working at high sampling rates. High-gain high-speed amplifiers are normally required to maximize the performance of ADCs. It will be discussed in more detail in Section 4.3.4.

Offset errors (comparator, sample-and-hold and amplifier) and sub-ADC nonlinearities are two other sources of error presented in the ADC, but through digital error correction, the effect of these errors can be reduced or eliminated [Lewis and Gray, 1987]. The basic idea of the digital error correction is that by reducing the interstage gain by $1/2$, the error caused by offsets and nonlinearities can be tolerated and finally corrected in the digital domain. Figure 3.10(a) shows an example of a digitally corrected pipeline ADC. If the first stage is perfectly linear, then only half of the conversion range of the second stage is used. When the nonlinearity in the first stage is in the range of $\pm 1/2$ LSB, it still does not cause over-range problem in the second stage. So one bit in the second stage is reserved to correct errors of the first stage, and the effective number of bits in the ADC is $(n_1 + n_2 - 1)$.

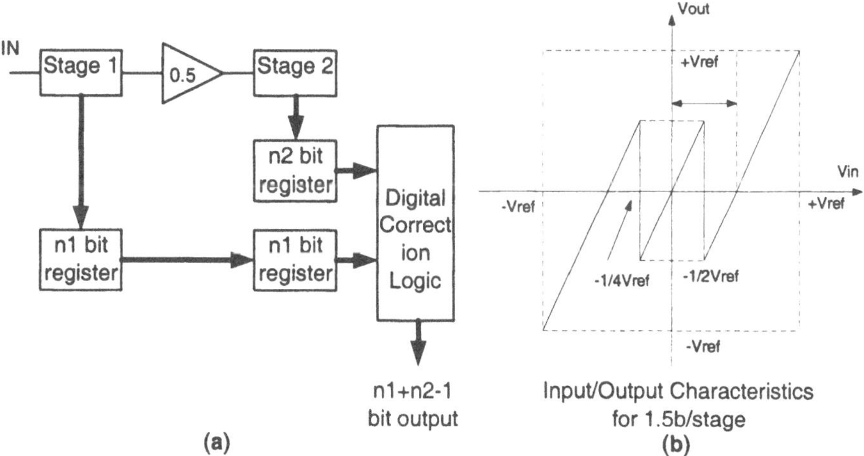

Figure 3.10. Digital Error Correction

Figure 3.10(b) illustrates the input/output transfer function of a commonly used 1.5b/stage architecture [Lewis and Gray, 1987]. In this architecture, each stage resolves two bits with a sub-ADC consisting of two comparators. The input signal range is from $-V_{ref}$ to $+V_{ref}$, and the two comparators have threshold voltages of $+V_{ref}/4$ and $-V_{ref}/4$. The sub-DAC has three output levels: $-V_{ref}$, 0, $+V_{ref}$. The input/output transfer function can be written as:

$$V_{out} = \begin{cases} 2V_{in} - V_{ref} & when V_{in} > V_{ref}/4 & D_{out} = 10 \\ 2V_{in} + V_{ref} & when V_{in} < -V_{ref}/4 & D_{out} = 00 \\ 2V_{in} & when - V_{ref}/4 < V_{in} < V_{ref}/4 & D_{out} = 01 \end{cases}$$

where D_{out} is the digital output codes from the sub-ADC. As can be seen from Figure 3.10(b), through the use of digital error corrections, offsets up to $\pm V_{ref}/4$ can be tolerated in this architecture.

In summary, digital error correction techniques relax the requirements on comparators, sample-and-hold circuits, and amplifiers. The cost is the increased number of comparators and some digital circuitry (which however are very simple).

4.3 Design Considerations

In this section, practical design considerations of low-power, high-speed, high-resolution pipeline ADCs are discussed.

4.3.1 Size of Capacitors

As discussed above, thermal noise (or kT/C noise) is a major source of error in a pipeline ADC. The only way to reduce thermal noise is to increase the size of sampling capacitors C, which in turn increases the power consumption of amplifiers associated with charging/discharging capacitors. Therefore there exists a tradeoff when choosing the capacitor size. Assume we have a N-bit ADC, with full scale input signal V_{FS} and sampling capacitor size C_s, then the SNR of the ADC when considering thermal noise and quantization noise only can be calculated as:

$$SNR = 10 \times \log \left[\frac{V_{FS}^2/2}{\frac{kT}{C_s} + \frac{(2V_{FS}/2^N)^2}{12}} \right] \tag{3.3}$$

In Figure 3.11, the maximum SNR versus different sampling capacitor values is plotted. In practice, minimum size capacitors for a given noise requirement should be chosen to reduce power consumptions.

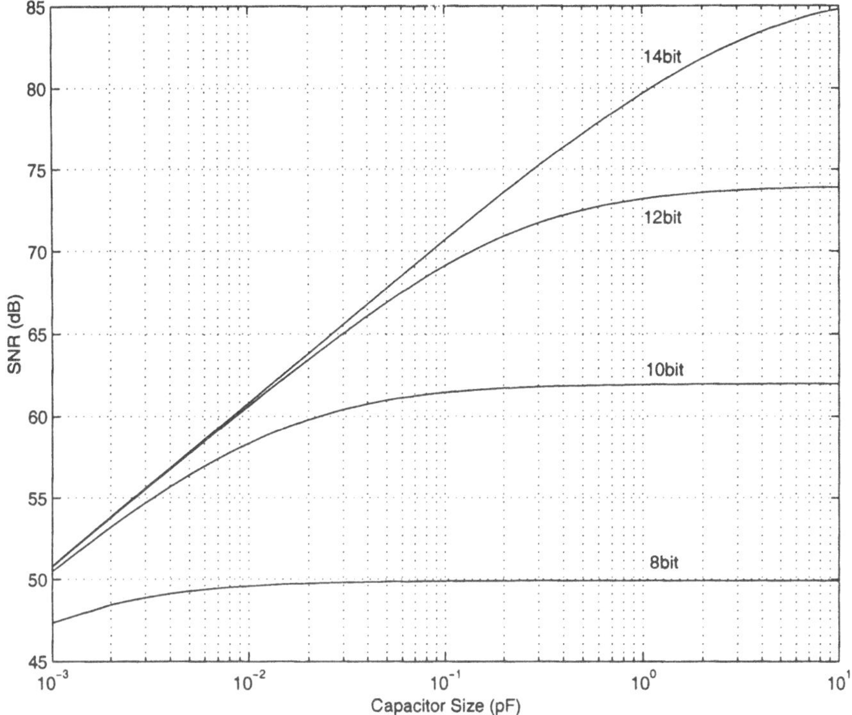

Figure 3.11. Maximum SNR versus Different C_s

4.3.2 Capacitor Matching

In pipeline ADCs, the sampling capacitors are associated with the gain amplifiers and the sub-DACs, so capacitor matching is another key factor when choosing the capacitor size. The matching requirement on the capacitors is determined by the allowable gain error and the required ADC accuracy. Shown in (3.4) and (3.5) are the relationships between input, reference and output voltages of the circuit shown in Figure 3.7:

$$V_{out} = \frac{(C_1 + C_f)}{C_f} V_1 - \frac{C_1}{C_f} V_{DAC} \tag{3.4}$$

$$\Delta V_{out} = \Delta(\frac{C_1 + C_f}{C_f}) V_I - \Delta(\frac{C_1}{C_f}) V_{DAC} \tag{3.5}$$

If the 1.5b/stage architecture described in last section is used, then the new transfer function can be simplified as:

$$V_{out} \approx (2 + \frac{\Delta C}{C})V_I - (1 + \frac{\Delta C}{C})V_{DAC} \quad (3.6)$$

Due to process gradient and random effects, capacitors laid out identically still show mismatch in value, and the overall untrimmed capacitors are limited to 8 $\sim$ 10-bit linearity depending on the CMOS technology and the capacitor size. Smaller capacitors tend to have worse matching, so the minimum kT/C-limited capacitor size may not be suitable for high resolution applications. In that case, certain trimming/calibration techniques can be employed to obtain higher resolutions.

4.3.3 Amplifier Architecture

Speed vs. power consumption is a key tradeoff in amplifier designs. Figure 3.12 and 3.13 illustrate several commonly used amplifier architectures. The single-stage amplifier (Figure 3.12) is a good choice for high-speed medium-resolution ADCs. To improve the DC-gain, the two-stage architecture with a Miller or cascode compensation can be used, but it tends to reduce the bandwidth and consume more power. Another way to increase the DC gain is to add a gain-boost amplifier to the main amplifier (Figure 3.13). However, the drawback is that it has less output swing than the two-stage architecture.

4.3.4 Amplifier Requirements

Finite gain and incomplete settling of amplifiers cause errors in pipeline ADCs. In this section, the requirements for amplifiers in high-speed high-resolution ADCs are analyzed.

1 DC-Gain Requirement:

For a finite opamp gain A, the relationship between the output and input of an interstage gain-amplifier can be written as:

$$V_{out} = \frac{1}{1 + A \cdot f}[(1 + \frac{C_s}{C_f})V_{in} - V_{DAC}] \quad (3.7)$$

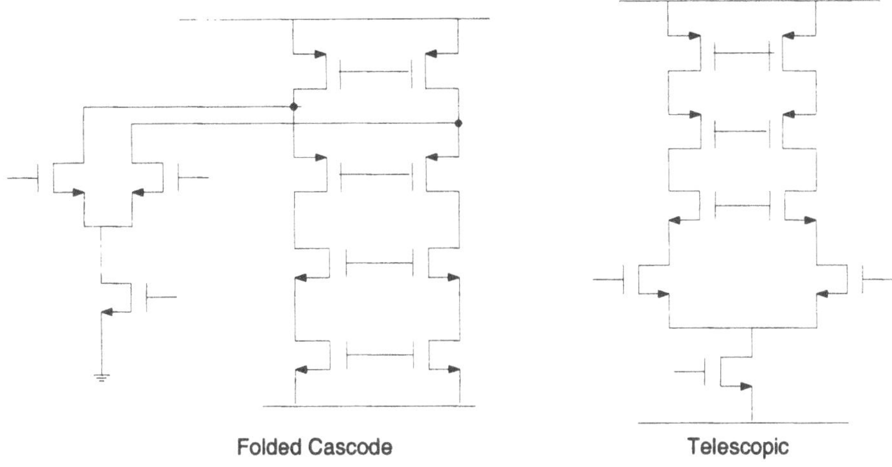

Figure 3.12. Amplifier Architecture 1 and 2

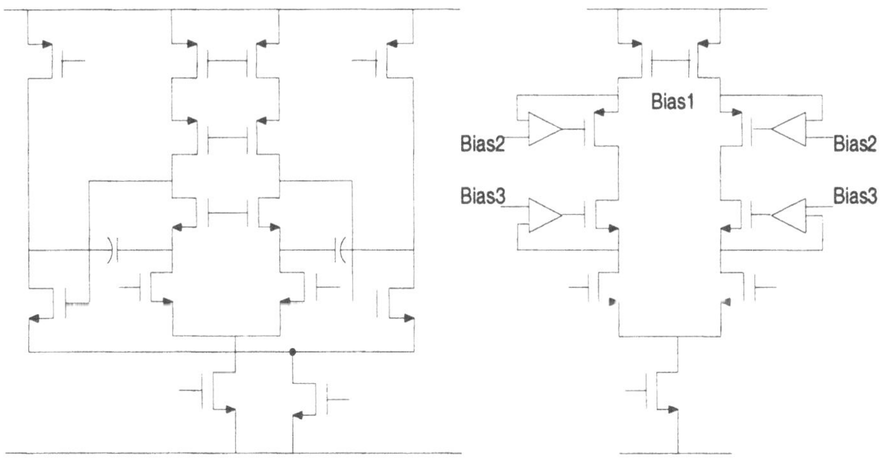

Figure 3.13. Amplifier Architecture 3 and 4

where $f = \frac{C_f}{C_f+C_s+C_{opamp}}$ is the feedback factor, and C_{opamp} is the input capacitance of the amplifier. If $A \cdot f \gg 1$, then the gain error of the interstage amplifier can be expressed as:

$$\frac{\Delta G}{G} \approx \frac{1}{A \cdot f} \tag{3.8}$$

For a N-bit ADC with a B-bit/stage architecture, the first stage gain error should be less than $1/2$ LSB of the full range of the second stage ($= V_{ref}/2^{N-B}$) to prevent any missing codes. Therefore A should be

$$A > 2 \times 2^{N-B} \times \frac{1}{f} > 2^{N+1} \tag{3.9}$$

This is the minimum requirement on A. In practice, the DC gain should be much larger than this value, since errors caused by other sources such as capacitor mismatches and incomplete amplifier settling are not included in (3.7).

2 Gain-Bandwidth Product Requirement:

The gain-bandwidth product (GBW) of the amplifier determines the maximum usable sampling rate in a pipeline ADC. If a single-pole amplifier model is assumed, the output voltage of the gain amplifier is given by:

$$V_{out}(t) = (1 - e^{-t/\tau})[(1 + \frac{C_s}{C_f})V_{in} - V_{DAC}] \tag{3.10}$$

where

$$\tau = \frac{C_L + C_{out} + \frac{C_f(C_s+C_{opamp})}{C_f+C_s+C_{opamp}}}{g_m} \cdot \frac{C_f + C_s + C_{opamp}}{C_f} \tag{3.11}$$

is the time constant for settling, C_L is the load capacitance, and C_{out} is the output capacitance of the amplifier. If $\frac{C_f(C_s+C_{opamp})}{C_f+C_s+C_{opamp}}$ is small compared to $C_L + C_{out}$, then the time constant can be further related to the GBW of the amplifier as

$$\tau \approx \frac{1}{2\pi \cdot GBW} \cdot \frac{1}{f} \tag{3.12}$$

The required time for the amplifier output to settle to within $1/2^N$ of the final value can be found as:

$$T_{settle} > N \cdot \ln(2.0) \cdot \tau \tag{3.13}$$

For pipeline ADCs, the allowed settling time is a little less than $1/2$ of a clock period. Then the minimum value for GBW of the amplifier should be

$$GBW \approx \frac{1}{2\pi \cdot f \cdot \tau} > \frac{N \cdot \ln(2)}{f \cdot \pi} \cdot F_s \tag{3.14}$$

where f is the feedback factor, and F_s is the sampling frequency.

3 Slew Rate:

Slew rate limited settling should be avoided in the amplifier design, since it is signal amplitude dependent and causes distortions in the output signal. Therefore, a sufficiently large value of opamp bias current should be used:

$$I_b > 2\pi \cdot F_{in} \cdot A_m \cdot C_{L,eff} \tag{3.15}$$

where F_{in} and A_m are the frequency and amplitude of the input signal, respectively, and $C_{L,eff} = C_L + C_{out} + \frac{C_f(C_s+C_{opamp})}{C_f+C_s+C_{opamp}}$ is the effective load capacitance of the amplifier.

4 Thermal Noise:

Two types of thermal noise should be considered in the pipeline ADC: the noise from the sampling switches, and the noise from the amplifier [Abo, 1999]. Therefore, the total noise spectrum density is

$$v_n^2 = \frac{1}{f}\frac{kT}{C_f} + v_{opamp}^2 \tag{3.16}$$

4.3.5 Error Tolerances

In the previous section, different errors caused by the capacitor mismatch, finite amplifier DC gain, incomplete amplifier settling, and noise are analyzed separately, and the requirements on capacitors and amplifiers are derived accordingly. In practice, the total error is the sum of all above errors. For a N-bit ADC with a B-bit first stage, the first stage output error can not be larger than $1/2$LSB at the $(N-B)$-bit level, therefore, the total error caused by all sources is limited by:

$$\Delta V_1 < \frac{1}{2}\frac{V_{ref}}{2^{N-B}} \tag{3.17}$$

The error limit of other stages can also be obtained using the same approach.

5. Power Optimization Techniques

In a pipeline ADC, the major power dissipation components are residue amplifiers and comparators, while the other components including clock drivers and digital circuits contribute a relatively small amount to the overall power dissipation. As a result, optimizing the power consumption of amplifiers and comparators will effectively reduce the total power of the ADC. In this section, several power optimization techniques are discussed.

5.1 Optimizing the Stage Resolution

The first choice a designer may meet when designing a pipeline ADC is the stage resolution. The number of bits per stage has a large impact on the speed, power, and accuracy requirements of each stage. On the one side, for fewer number of bits per stage, more pipeline stages are required, which means more amplifiers; on the other side, the comparator requirements are more relaxed, and the bandwidth of interstage gain amplifiers is larger if a lower stage resolution is used. The optimum value of the stage resolution is dependent on the overall ADC specifications.

For high-speed pipeline ADCs with $8 \sim 10$-bit resolutions, minimizing the stage resolution reduces both the chip area and power consumption due to two reasons. The first reason is that the bandwidth of a gain amplifier depends on the interstage gain. With a minimum gain of 2, the load capacitor is minimized and the feedback factor is maximized, thus a large amplifier bandwidth can be achieved. As a consequence, for the same speed requirement, the biasing current of the amplifier can be reduced, resulting in large power savings.

The second reason is that by using the minimum stage resolution, large comparator offsets in the sub-ADC section can be tolerated by using digital correction techniques. For example, by using the 1.5bit/stage architecture, only two comparators are needed in the sub-ADC, and comparator offsets up to $\pm V_{ref}/4$ can be corrected and will not affect the overall linearity or SNR of the ADC. This large tolerable offset range can simplify the comparator design, since it eliminates the need for low-offset pre-amp stages which dissipate DC power, instead, a dynamic comparator dissipating only a small amount of AC power can be used.

In above discussions, a certain relation between the power consumption of amplifiers and comparators is assumed [Mehr and Singer, 1999], and each stage has the same resolution. For resolutions greater than 10-bits, this architecture may not necessarily lead to minimum power consumptions. In that case, if the resolution of the first stage is increased, the accuracy requirements on subsequent stages will be greatly reduced. As a result, the size of amplifiers in those stages can be scaled down, hence a large power saving can be obtained.

Optimizing the stage resolution of very high resolution ADCs has recently become a topic of active research. However, it is beyond the scope of this work.

5.2 Dynamic Comparator

As mentioned above, the dynamic comparator without pre-amp stages can be used in low stage resolution ADCs where large comparator offsets can be tolerated. One example of such comparators [Cho and Gray, 1995] is shown in Figure 3.14. It consumes zero DC power, and the switching threshold is self-adjusted by the transistor widths ($W2$ and $W1$).

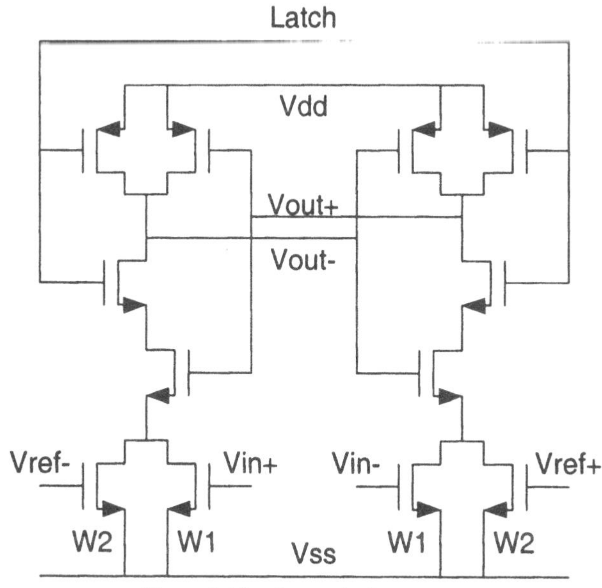

Figure 3.14. Dynamic Comparator

In Figure 3.14, the lower set of transistors operate in the triode region and acting as voltage-controlled resistors. The switching threshold of the dynamic comparator is set by the ratio of transistor widths $W1$ and $W2$, as given in (3.18):

$$V_{threshold} = \frac{W_2}{W_1} \cdot V_{ref} = \frac{W_2}{W_1} \cdot (V_{refp} - V_{refn}) \qquad (3.18)$$

It is important to note that this equation is valid only when the layout and the load capacitances are balanced. In practice, careful design and symmetrical layout should be taken to reduce the offset due to mismatch and process variations.

5.3 Capacitor/Amplifier Scaling

Another commonly used method to reduce power consumptions is through capacitor scaling. As discussed before, thermal noise as well as the capacitor matching sets the limit on the minimum usable sampling capacitor in a pipeline ADC. For the sample-and-hold and interstage gain amplifiers, the power consumption is dominated by the amplifiers charging/discharging capacitors. Therefore the kT/C-limited minimum size capacitors should be used to reduce power consumptions.

In pipeline ADCs, the accuracy requirements of later stages become more relaxed along the pipeline. For example, in a 10-bit ADC with 1.5-bit per stage, the first stage needs to have a 10-bit accuracy, while the N-th stage only needs $(10 - N + 1)$-bit accuracy. Therefore the later stages in the pipeline can be scaled down by using smaller amplifiers and sampling capacitors.

Optimizing the capacitor sizes is a complex task. Several practical factors such as parasitic capacitance, matching and process variation should all be taken into considerations.

5.4 Dynamic Biasing

Another power optimization technique for pipeline ADCs named dynamic biasing is proposed by author in [Shi et al., 2001b]. The basic idea is as follows. Take the simple configuration of an interstage gain amplifier as an example (Figure 3.15). During the sampling phase ϕ_1, the input signal charges the sampling capacitor C_s and feedback capacitor C_f, while at holding phase ϕ_2, charge redistribution causes the amplifier output to be $(1 + C_s/C_f)V_{in}$. Notice that the amplifier only operates at the high-gain state during phase ϕ_2, while during phase ϕ_1 it stays in the idle state. If we turn off the amplifier during ϕ_1, the function of the circuit will remain the same, but the total power consumption can be reduced by up to 30%.

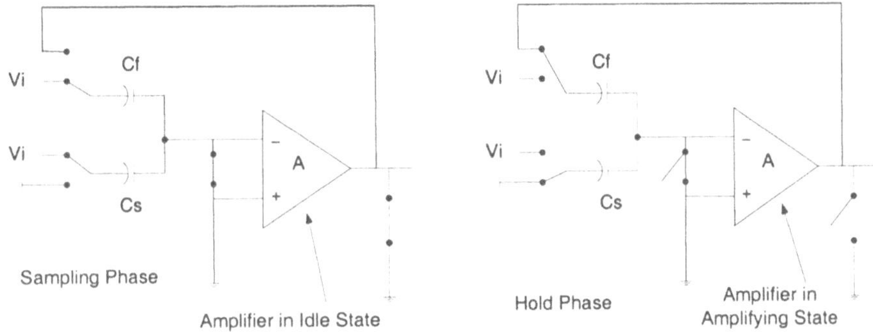

Figure 3.15. Basic Idea of Dynamic Biasing

This idea is similar to the switched-opamp technique first introduced by Jan Crols [Crols and Steyaert, 1994], but is more suitable for high speed applications. The underlying idea of the switched-opamp technique is to replace those switches in a switched-capacitor circuit which are not connected to constant reference voltages with a switchable opamp [Crols and Steyaert, 1994]. It is a promising way to design good-performance switched-capacitor circuits at very low supply voltages. But its major drawback is that it is not suitable for high-speed applications since the outputs of the switched-opamps are typically pulled to one of the supply rails during the idle phase, causing a long turn-on recovery time. The speed of switched-opamp circuits is usually limited within 10MHz. Dynamic biasing, on the other hand, does not turn off the opamp completely, instead, it leaves the biasing circuits unchanged, but partially turns off the supply current in the core part of the amplifier. Therefore, it is capable of operating at higher sampling frequencies.

The concept of the dynamic biasing will become more clear from the following design example.

Shown in Figure 3.16 is the core of a telescopic OTA, which can be used as an interstage gain amplifier as discussed in Section 4.3. The tail current is provided by two biasing transistors M_{b1} and M_{b2}, with their gates connected to the common-mode feedback circuit and the biasing circuit, respectively. A switch (M_{s1}) is added to the biasing circuit, which is controlled by a signal V_{ctrl}. During phase ϕ_2, M_{s1} is off, and the OTA functions as a gain amplifier; during phase ϕ_1, the OTA is in idle state (reset), so M_{s1} is on, partially turning off the OTA. Notice that in the reset state, the outputs of OTA are connected to the common-mode reference voltage, so turning off M_{b2} will not increase the current flowing through M_{b1} considerably due to the common-mode feedback.

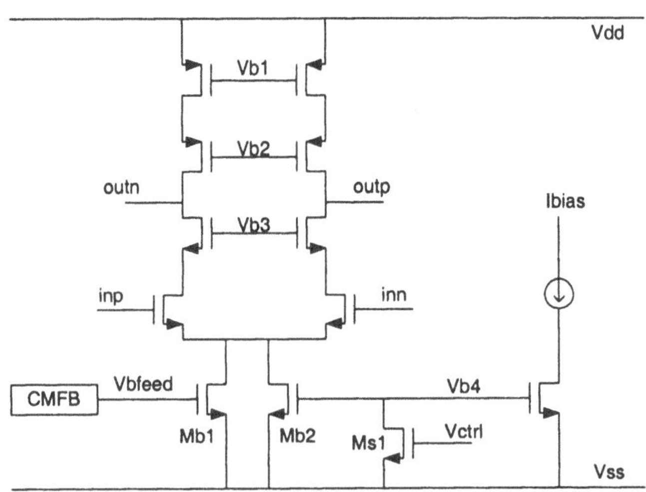

Figure 3.16. A Simplified Example of Dynamically Biased OTA

Two approaches can be employed to minimize the OTA turn-on recovery effect. First, all the biasing circuits are kept on with the exception of M_{b2}, therefore V_{b1}, V_{b2} and V_{b3} are kept constant during both phases, so the turn-on time for the OTA core is minimized. Second, the control switch M_{s1} can be turned off before the ϕ_1 to ϕ_2 transition, making the OTA return to normal biasing conditions before it enters the amplifying phase. An example of the control signal timing diagram is illustrated in Figure 3.17.

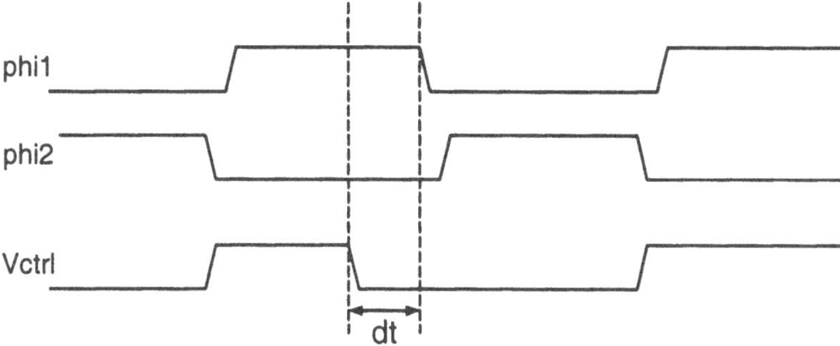

Figure 3.17. Timing Diagram of ϕ_1, ϕ_2 and V_{ctrl}

In summary, the existence of an idle state in the pipeline interstage gain amplifier makes the dynamic biasing a practical way to save power consumptions.

6. Summary

In this chapter, several high-speed ADC architectures are briefly reviewed, among which pipeline ADC is most suitable for low-power high-speed medium-to-high resolution applications. Several design issues of pipeline ADCs are then discussed, with the focus on the power-optimization techniques. A novel technique named dynamic biasing is proposed to reduce the total power consumption of pipeline ADCs by dynamically adjusting the biasing current of amplifiers.

Chapter 4

PROTOTYPE DESIGN: ADC FOR WLAN(DSSS)/WCDMA

1. Introduction

In this chapter, a prototype pipeline ADC design is described. The main objective of this design is to verify the feasibility of the dynamic biasing scheme in high-speed ADCs and study the effect of power optimization techniques. This ADC is one of the experimental chip-sets for WLAN(DSSS)/WCDMA direct-conversion receivers, and can be integrated with other baseband circuits into a single chip.

2. Applications

The prototype ADC is targeted for wideband wireless applications such as WLAN (DSSS) and WCDMA. Wireless local area network (WLAN) with direct sequence spread spectrum (DSSS), as defined in IEEE 802.11 standard, provides short range wireless connections with high data rates among computers, printers, and other network nodes. The high speed short range wireless systems can be used in environments such as offices, airports, hospitals, etc, where users are highly mobile. They operate at unlicensed Industry-Scientific-Medical (ISM) frequency bands. The date rates are 1Mbps or 2Mbps for typical systems and up to 11Mbps for high rate (DR) systems (Table 4.1).

WCDMA (Wideband Code Division Multiple Access) is the radio access technology selected by ETSI (European Telecommunications Standards Institute) in January 1998 for wideband radio access to support the third generation cellular services [Gee, 2001]. This technology provides data rates up to 2Mbps in the local area and 384Kbps in the wide area with full mobility. The nominal channel bandwidth is 5MHz, while higher bandwidthes of 10, 15, and 20MHz have been proposed to support higher data rates. With the emergence of WCDMA systems, multimedia services such as voice, internet access, and video-conference will turn into reality.

As shown in Table 4.1, WLAN(DSSS) and WCDMA have many similarities in terms of bandwidth, data rate, access scheme, etc. A zero-IF direct conversion receiver is suitable for both applications. From system perspective, the ADC in such a receiver has similar specifications when used in WLAN(DSSS) or WCDMA systems.

Table 4.1. Summary of WLAN (DSSS) and WCDMA Standards

Standard	Access Scheme	Frequency Range	Channel Spacing	Modulation Scheme	Data Rate
WLAN(DSSS)	CDMA	2.4GHz	up to 20MHz	QPSK	1~11Mbps
WCDMA	CDMA	2GHz	5MHz	QPSK	2Mbps

A single ADC can thus be designed for these two applications. A 7-bit 64MS/s pipeline ADC meets the requirements of both WLAN(DSSS) and WCDMA standards. The design of this ADC is described in the following sections.

3. Architecture

The 1.5b/stage architecture described in Chapter 3, Section 4.2 is chosen in this design since it has been shown to be effective in achieving high speed at low power [Abo and Gray, 1999]. As shown in Figure 4.1, the prototype ADC consists of five identical pipeline stages in which each stage resolves one effective bit while keeping 0.5 redundant bit for digital error corrections, followed by a 2-bit flash stage. The 12-bit digital codes from six stages are passed to the digital correction circuit where the 7 effective bits are generated.

The switched-capacitor implementation of each pipeline stage is shown in Figure 4.2. It consists of a 1.5-bit sub-ADC, a 1.5-bit sub-DAC and an

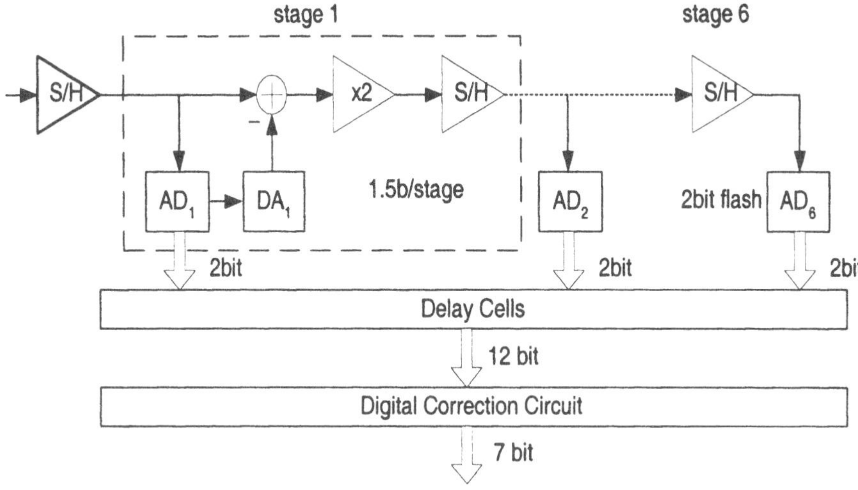

Figure 4.1. Block Diagram of the 1.5b/stage Pipeline ADC

interstage amplifier with gain of 2. This circuit operates at a two-phase clock. During the sampling phase (ϕ_1), the input signal is simultaneously sampled by the 1.5-bit sub-ADC and two capacitors C_s and C_f (C_s=C_f). The 1.5-bit sub-ADC consisting of two comparators then produces two digital outputs (D_1 and D_0). During the holding phase (ϕ_2), the bottom plate of C_s is connected to one of the reference voltages (V_{refp} or V_{refn}) or the common mode voltage (V_{cm}), depending on the output of the 1.5-bit sub-DAC (X, Y, or Z), while at the same time, the bottom plate of C_f is connected to the output of the gain amplifier. A residue voltage is thus generated at the amplifier output as given by:

$$V_{out} = \begin{cases} (1 + C_s/C_f) \cdot V_{in} - C_s/C_f \cdot V_{ref} & if V_{in} > V_{ref}/4 \\ (1 + C_s/C_f) \cdot V_{in} + C_s/C_f \cdot V_{ref} & if V_{in} < -V_{ref}/4 \\ 2C_s/C_f \cdot V_{in} & if - V_{ref}/4 < V_{in} < V_{ref}/4 \end{cases}$$

Note that a two-phase non-overlapping clock is used, and the “bottom-plate sampling" is implemented by turning off switches connecting to the analog ground before those connecting to an analog input [Kim et al., 1997]. In this way, the errors caused by capacitor parasitics and signal-dependent charge injections can be greatly reduced.

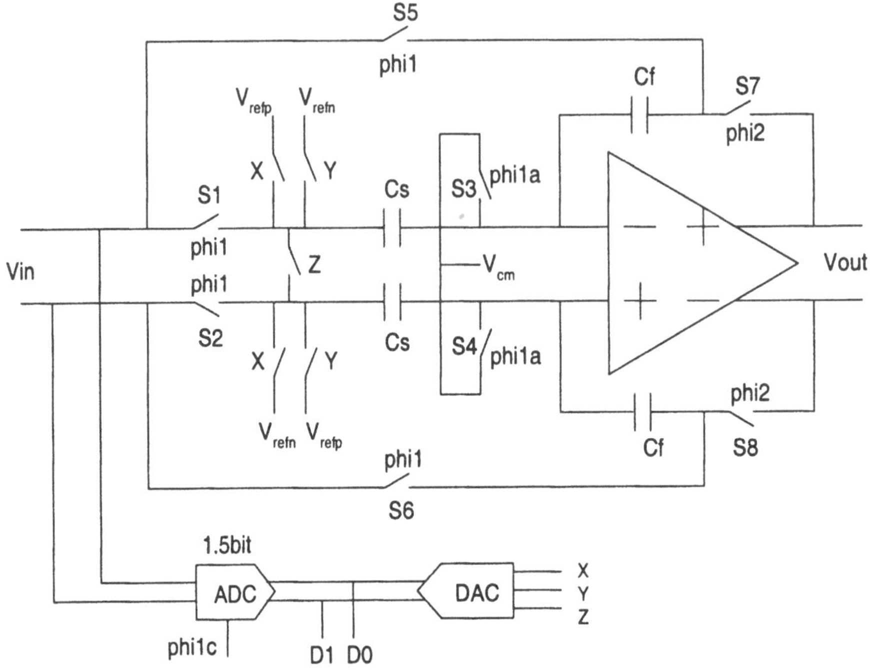

Figure 4.2. A 1.5b Pipeline Stage

4. High-Speed OTA

A transconductance amplifier (OTA) is typically employed as the gain amplifier in pipeline ADCs. In this section, the design of a high-speed medium-gain OTA is described in details.

4.1 OTA Requirements

The ADC resolution determines the required OTA signal swing and open loop DC gain, while the sampling rate determines the OTA's gain-bandwidth product. Equation (3.9) gives the minimum requirement on the open loop DC gain of the OTA. In practice, the gain needs to be two or three times of this theoretical value to compensate other error sources and tolerate any process variations, implying that an open loop DC gain larger than 50dB is required in this design. For the gain bandwidth requirement, Equation (3.14) should be revisited if the dynamic biasing is employed. In that case, T_{settle} is no longer half of a clock period ($\frac{1}{2F_s}$), instead it should be $\frac{1}{3F_s} \sim \frac{1}{4F_s}$ because of less available settling time. As a result, the gain bandwidth requirement is modified to be

$$GBW \approx \frac{1}{2\pi \cdot f \cdot \tau} > \frac{2 \cdot N \cdot \ln(2)}{f \cdot \pi} \cdot F_s \tag{4.1}$$

which equals 400MHz in this design.

4.2 OTA Topology

From above discussions, an OTA with a high GBW and a medium open loop DC gain is required. A single stage non-folded-cascode (telescopic) OTA is chosen for several reasons. First, the medium gain requirement makes it possible to use a single-stage architecture. Second, the speed and simplicity of having only NMOS devices in the signal path renders it superior to its folded-cascode counterpart [Conroy et al., 1993]. The last but most important reason is that the telescopic OTA has lower power consumption than its folded-cascode counterpart, since it has only two current legs instead of four.

Figure 4.3 depicts the fully differential OTA used in the design (excluding biasing and CMFB circuits).

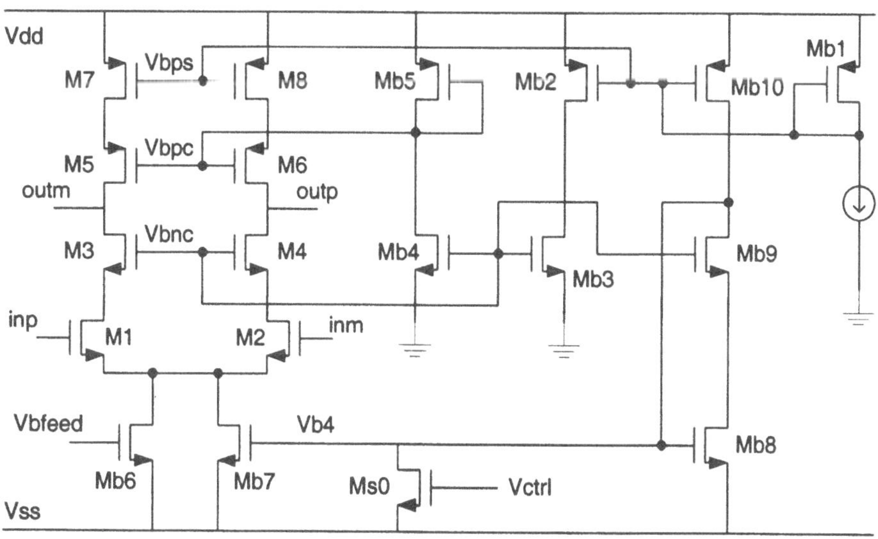

Figure 4.3. A Simple Telescopic OTA

4.2.1 DC Gain

The open loop DC gain of this OTA can be written as:

$$A_{DC} = g_{m1} \cdot [g_{m3}r_{o3}r_{o1}]//[g_{m5}r_{o5}r_{o7}] \tag{4.2}$$

To increase A_{DC}, the PMOS load devices M_7 and M_8 are made long channels to maximize the output resistance. This will not slow down the OTA since those devices are not in the signal path.

4.2.2 Gain-Bandwidth

The GBW of the OTA in closed-loop can be expressed as

$$GBW = \frac{g_{m1} \cdot \beta}{2\pi C_{L,eff}} \tag{4.3}$$

where

$$C_{L,eff} = C_L + C_{out} + \frac{C_f(C_s + C_{opamp})}{C_f + C_s + C_{opamp}} \tag{4.4}$$

is the effective load capacitance of the OTA, and $\beta = C_f/(C_f + C_s + C_{opamp})$ is the feedback factor. Substituting the expression for β into (4.3), the closed-loop bandwidth is

$$GBW = \frac{g_{m1}}{2\pi[C_s + C_{opamp} + \frac{(C_L+C_{out})(C_f+C_s+C_{opamp})}{C_f}]} \tag{4.5}$$

Since the f_T of NMOS devices are approximately three times that of PMOS devices, the input devices of this telescopic amplifier are chosen to be NMOS. Minimal length devices are also used in the signal path to maximize the GBW of the OTA. Of course, this will increase the flicker noise in the OTA, which however does not pose a major problem in this circuit.

4.2.3 Slew Rate

The slew rate of this OTA is given by

$$SR = \frac{I_{Mb6} + I_{Mb7}}{C_{L,eff}} \tag{4.6}$$

4.2.4 Thermal Noise

In the telescopic OTA, there are only 4 noise contributing devices (M_1, M_2, M_7, and M_8), since the noise of cascode devices (M_3, M_4, M_5, and M_6) are negligible. Therefore, the total input referred thermal noise power spectrum of the OTA can be calculated as:

$$v_{in,thermal}^2(\Delta f) = 2 \cdot 4kT \cdot \frac{2}{3} \cdot (\frac{1}{g_{m1,2}} + \frac{g_{m7,8}}{g_{m1,2}^2}) \cdot \Delta f \tag{4.7}$$

To decrease noise, it is desired to increase $g_{m1,2}$ and decrease $g_{m7,8}$.

4.3 CMFB

In fully-differential OTAs, common-mode feedback (CMFB) is required to define the common-mode voltages at the high-impedance output nodes. A switched-capacitor CMFB is preferred in SC circuits, because it allows a larger output signal swing and dissipates no DC power [Jones and Martin, 1996]. Figure 4.4 illustrates the CMFB circuit adopted for this OTA. It operates at a two phase non-overlapping clock, and V_{cm} is the desired output common-mode voltage. Two capacitors labeled C_f sense the output voltages (V_{outp}, V_{outm}) and generate the average voltage V_{bfeed}, which is used to control the OTA current source device M_{b6}. The DC voltage across C_f is determined by capacitors C_r which are switched between bias voltages (V_{b4}, V_{cm}) and between being in parallel with C_f. The capacitors C_r are chosen to be $1/5$ the sizes of C_f. Minimum sizes devices are used for the CMOS switches.

It is also important to note that the GBW of common-mode loop should be high enough to allow for fast recovery when it is used in the dynamic biasing configuration.

4.4 Results

Simulation shows that the designed OTA has a $GBW > 700MHz$, phase margin $> 59^o$, open loop gain $> 59dB$, slew rate $> 500V/\mu S$ with 1.5pF load capacitance. It dissipates 3mA current with a single 3V power supply. This same OTA is used in both sample-and-hold and the first pipeline stage.

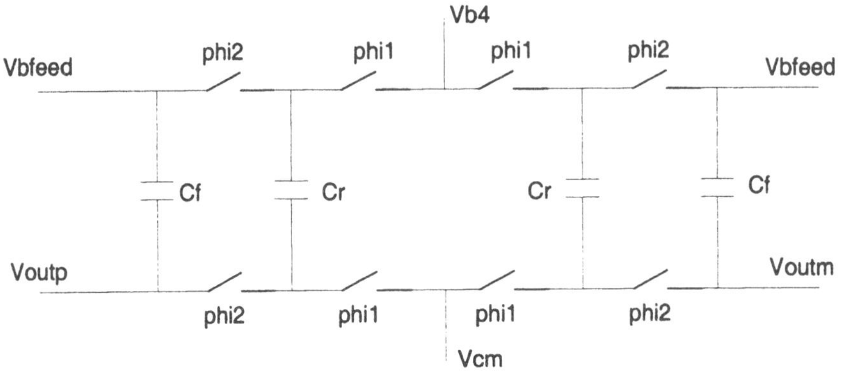

Figure 4.4. Switched-Capacitor Common-Mode Feedback Circuit

5. Comparator

As discussed in Chapter 3, by using digital correction and minimum number of bits per stage, comparator offsets up to $\pm V_{ref}/4$ can be tolerated without causing code errors, therefore simple dynamic comparators can be used to save powers. The comparator introduced in Figure 3.14 is used in this design. A digital latch is added after the comparator to synchronize the outputs.

In conventional pipeline ADCs, a two phase non-overlapping clock (ϕ_1, ϕ_2) is used as illustrated in Figure 4.5, where t_{nov} is the non-overlapping interval during which neither phase is active. This time is used for the comparators to digitize the sample and the sub-DAC to select a level from $+V_{ref}$, $-V_{ref}$, or V_{cm}. There is a tradeoff when choosing the value of t_{nov}. If t_{nov} is too small, the comparator, sub-DAC and the digital logics will not have enough time to react, and the probability of meta-stability in the comparator will increase. On the other hand, if t_{nov} is too large, then the OTA will have less time to settle. This is one of major problems in high-speed pipeline ADC designs.

A three-phase non-overlapping clock is proposed here to solve the above problem. Shown in Figure 4.5 is the timing diagram of this new clock. The third phase labeled phi_{comp} is added to control the comparators, which has a longer non-overlapping interval $t_{nov,comp}$ than t_{nov}. In this configuration, the comparator has longer time to compare and digitize the signal, while the OTA still has the same duration to settle. The underlying idea of this approach is that because of using digital correction, the comparator can sample the signal before the previous stage is fully settled. It will not cause any code errors as long as the settling error is much smaller than the correction range $V_{ref}/4$.

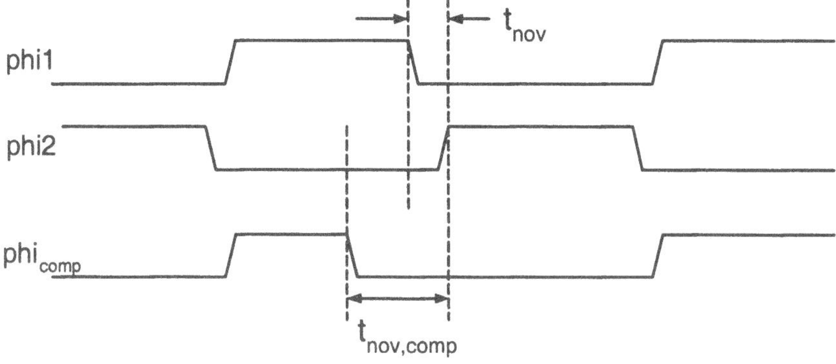

Figure 4.5. Timing Diagram for Comparator

This novel clock is helpful for increasing the maximum sampling rate in high-speed pipeline ADCs.

6. Clock Generator

Based on above discussions, a three-phase non-overlapping clock is employed in this ADC. The clock generator circuit is shown in Figure 4.6. The non-overlapping intervals t_{nov} and $t_{nov,comp}$ are determined by the propagation delays in inverters, NAND and NOR gates.

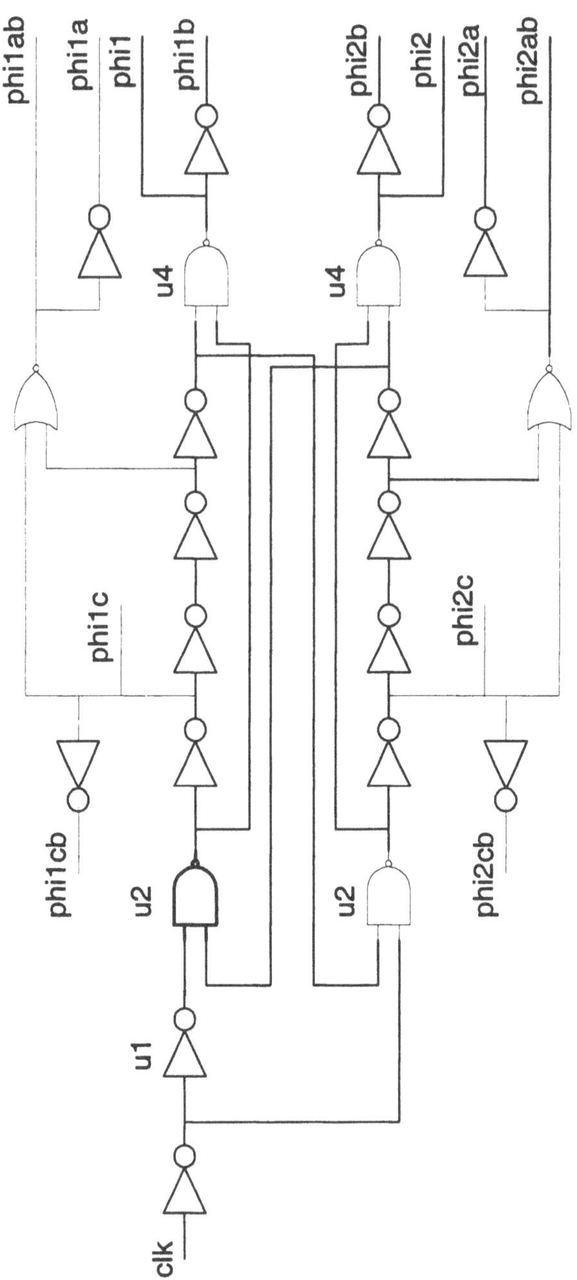

Figure 4.6. Clock Generator

7. Smart-Biasing Technique

Notice that for a pipeline ADC, it is not necessary to keep it working at the maximum sampling frequency all the time, instead, the sampling rate can be adjusted based on the input signal level and frequency range. A novel technique named smart-biasing is proposed here for further power savings. The basic idea is that several different biasing currents for the OTA can be selected based on different sampling frequency ranges. The step-size can be non-uniform. For example, at low sampling frequencies, each 5MHz frequency range may correspond to a biasing current, while at high frequencies, the step size can be larger (say, 10MHz, or 20MHz). The sampling frequency and biasing current can be easily controlled by digital microprocessors. In this way, one ADC can be power-efficiently used for different frequency bands.

8. Prototype Implementation

The prototype ADC was fabricated using HP $0.5\mu m$ CMOS process through MOSIS. The process has standard threshold voltage levels of 0.7V and 0.9V for NMOS and PMOS devices respectively. Linear capacitors were implemented using poly over n+ diffusion in a n-well.

A chip photograph of the prototype ADC is shown in Figure 4.7. The size of the active area is $1.2mm \times 1.1mm$.

9. Performance of the Prototype ADC

This ADC is simulated at the transistor level using BSIM3v3 model. As introduced in Chapter 3, SFDR (spurious-free dynamic range) and SNDR (signal-to-noise-plus-distortion ratio) are usually used to characterize the dynamic performance of the ADC. FFT (Fast Fourier Transform) is employed to obtain the power spectrum of the ADC, and Hanning window is chosen to reduce the leakage caused by finite data length. Figure 4.8 depicts the simulated power spectrum of the ADC output with 2MHz input signal sampled at 64MS/s. The SFDR is found to be around 44dB, which is the ratio of the signal power to the third order harmonic distortion power. The SNDR is calculated to be 39dB, which is the ratio of signal power to total noise plus distortion power. The ADC performance basically reaches our design goal.

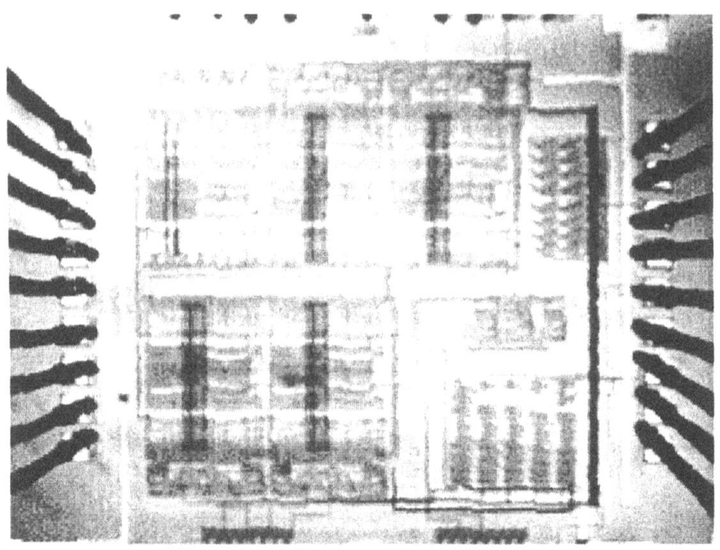

Figure 4.7. Chip Photograph of the Experimental Pipeline ADC

Table 4.2. Performance Summary of the Pipeline ADC

Parameter	Values
Technology	$0.5\mu m$ CMOS
Resolution	7 bits
Conversion Rate	64 MS/s
SFDR	44 dB
SNDR	39 dB
Active Area	1.2mm x 1.1mm
Supply Voltage	3.0 V
Power Consumption	31 mW

This chip operates at a single 3V voltage supply, and the power consumption is 31mW (excluding digital output driver).

The performance of this prototype ADC is summarized in Table 4.2.

10. Summary

One single ADC can be used for both WLAN (DSSS) and WCDMA applications. An experimental prototype ADC was designed and imple-

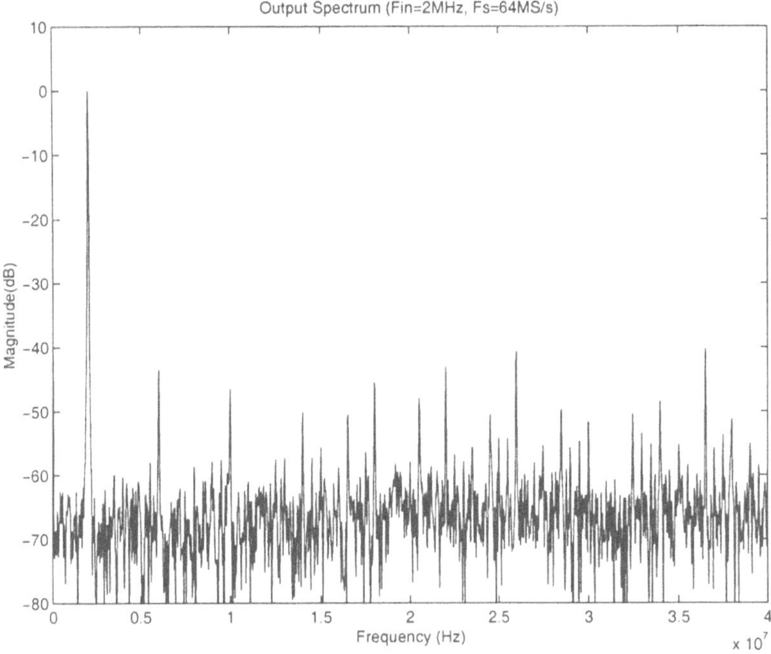

Figure 4.8. Power Spectrum of the ADC Output

mented in a $0.5\mu m$ CMOS technology. This pipeline ADC has a 7-bit resolution, and up to 64MS/s sampling rate. It operates at a single 3V supply voltage, dissipates only 31mW power.

Chapter 5

DESIGN CONSIDERATIONS OF LOW VOLTAGE ADCS

There has been an increasing need for the implementation of analog CMOS circuits that operate at very low supply voltages (1.0V to 1.8V). The switched-opamp technique is a variation of conventional switched-capacitor circuits and is capable of operating at very low supply voltages without the need of on-chip voltage multiplier or low-threshold devices. In this chapter, design considerations of low voltage ADCs are discussed and a modified switched-opamp technique is proposed.

1. Introduction

Besides low power consumption, low supply voltage is perhaps the most important design constraint in the past years. This trend has been driven by three factors. The first is the fast expanding market of portable and battery operated electronic devices, such as personal communication devices, consumer electronics, etc. Low voltage operation is desired because it requires less batteries, which in turn reduces the size and weight of the system. The second is the continuous scaling of CMOS technology, which is driven by digital system needs to enhance the circuit speed performance and increase the integration density by continuously reducing the channel length. Lower supply voltages are required for shorter channel lengths. According to the Semiconductor Industry Association's roadmap [roa, 1999], the maximum supply voltage of digital circuits will drop to 1.5V by 2001 and to 0.9V by 2005 (Table 5.1). The third factor is the trend for System-on-Chip (SoC) design. It makes it mandatory for analog designer to use the latest digital CMOS technology in order to integrate the entire system on a single chip.

Table 5.1. SIA Technology Roadmap 1999

Year	2001	2003	2005	2008	2011	2014
Gate length (μm)	0.15	0.12	0.10	0.07	0.05	0.035
Supply Voltage (V)	1.2-1.5	1.2-1.5	0.9-1.2	0.5-0.9	0.5-0.6	0.3-0.6

The ADC is an important system building block providing the interface between the analog world and the digital signal processing (DSP) circuit. It is beneficial and often a system requirement to integrate the ADC block with the digital system. Therefore there has recently been an increasing demand for low voltage ADCs.

2. Challenges in Low Voltage ADC Design

In digital circuits, the power consumption is a strong function of the supply voltage:

$$P \propto CV_{dd}^2 f \tag{5.1}$$

where C is the load capacitance, f is the average switching frequency and V_{dd} is the supply voltage. It is obvious that lowering the supply voltage directly reduces the power consumption in digital circuits.

However, this rule does not apply to analog/mixed-signal circuits. On the contrary, the power consumption in analog circuits will increase as the supply voltage decreases in order to keep the same performance including speed, signal-to-noise ratio, etc [Sansen et al., 1998]. This can be explained as follows. In analog circuits, the available signal range decreases when the supply voltage is scaled down, then the dynamic range becomes an important issue. In order to maintain the same dynamic range with reduced signal voltages, the noise in the circuits must also be reduced accordingly. In typical switched-capacitor circuits, this requires larger capacitor values, which in turn leads to a larger power consumption. The relationship between power consumption and supply voltage in analog circuits can be qualitatively given as [Abo, 1999]:

$$P \propto kT \cdot DR \cdot (\frac{V_{gt}}{\alpha^2 V_{dd}}) \cdot f_s \tag{5.2}$$

where DR is the dynamic range, $V_{gt} = V_{gs} - V_t$ is the gate over-drive voltage, αV_{dd} is the available signal swing, and f_s is the sampling frequency. It shows that the power is inversely proportional to the supply voltage in analog circuits.

Moreover, the reduction of the supply voltage greatly complicates the design of analog building blocks at the transistor level. Besides the dynamic range issue as mentioned above, there are limited number of circuit topologies suitable for low voltage operations. Given the typical switched-capacitor S/H shown in Figure 5.1 as an example. It consists of three basic elements: capacitors, switches, and opamps. The properties of capacitors in CMOS technologies are not strongly affected by the supply voltage reduction. However, opamps become more difficult to design at low voltages. It is not practical to stack more than two transistors in the signal path when the supply voltage drops below 1.8V, therefore in order to achieve enough DC gain, two-stage or multi-stage opamps are usually needed, which tends to consume more power with limited bandwidth.

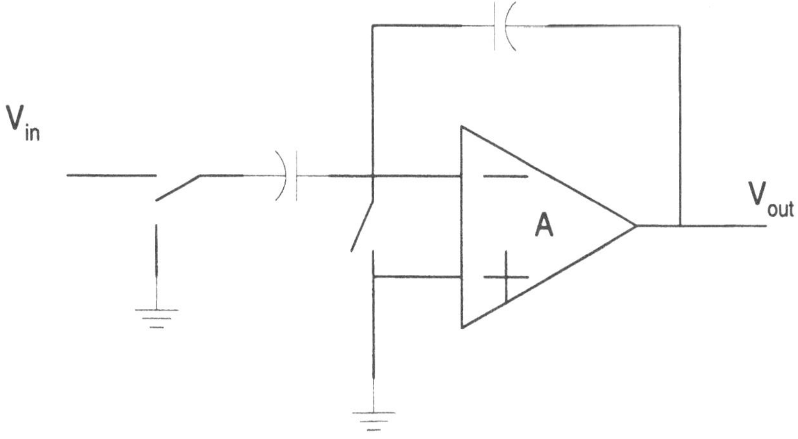

Figure 5.1. A Typical Switched-Capacitor S/H

Figure 5.2 shows the simplest input and output stage of an opamp. The minimum input common mode voltage is given by:

$$V_{in,cm} \geq V_{ds,sat} + V_{T,n} + V_{in,swing}/2 \tag{5.3}$$

where $V_{T,n}$ is the threshold voltage of NMOS devices, $V_{ds,sat}$ is the saturation voltage, and $V_{in,swing}$ is the input signal swing (a small value due to virtual ground principle).

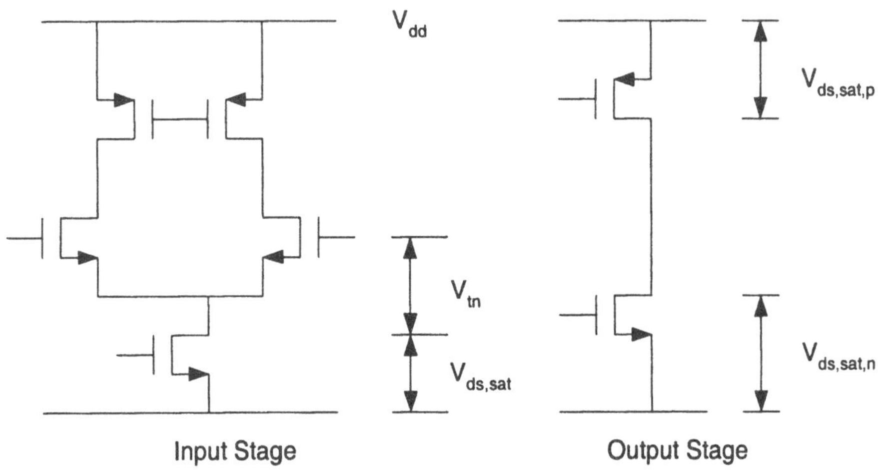

Figure 5.2. DC Operating Point Requirement for the Input and Output Stage of an OPAMP

In the output stage, the minimum power supply voltage is given by:

$$V_{dd} \geq V_{ds,sat,n} + V_{ds,sat,p} + V_{out,swing} \tag{5.4}$$

where $V_{out,swing}$ is the output signal swing.

2.1 Low Voltage CMOS Switches

Another challenge in low voltage analog design is the difficulty of implementing MOS switches. A typical switched-capacitor switch consists of complementary driven NMOS and PMOS transistors as shown in Figure 5.3(a). When the input signal V_{in} is in the range of $[0, V_{T,p}]$, only NMOS transistor conducts; when V_{in} is in the range of $[V_{dd} - V_{T,n}, V_{dd}]$, only PMOS transistor conducts. Both transistors are turned on when $V_{T,p} < V_{in} < V_{dd} - V_{T,n}$ ($V_{T,n}$ and $V_{T,p}$ are the threshold voltages of NMOS and PMOS devices, respectively).

Shown in Figure 5.3(b) is the switch conductance g_{ds} versus input signal V_{in}. The dashed line shows the individual conductances of the NMOS and PMOS transistors which are given by:

$$g_{ds,n} = \mu_n C_{ox}(\frac{W}{L})(V_{dd} - V_{in} - V_{T,n}) \tag{5.5}$$

and

$$g_{ds,p} = \mu_p C_{ox}(\frac{W}{L})(V_{in} - V_{T,p}) \tag{5.6}$$

respectively for square-law devices that operate in the linear region, where μ_n and μ_p are the electron and hole mobility in the channel, C_{ox} is the gate oxide capacitance per unit area, W and L are the effective width and length of the device, respectively. The solid line shows the effective parallel conductance. As can be seen from the figure, when $V_{dd} < V_{T,n} + V_{T,p}$, there exists an input signal range where none of the transistors conducts. So $(V_{T,n} + V_{T,p})$ is the fundamental limit of the minimum supply voltage for rail-to-rail operation in switched-capacitor circuits.

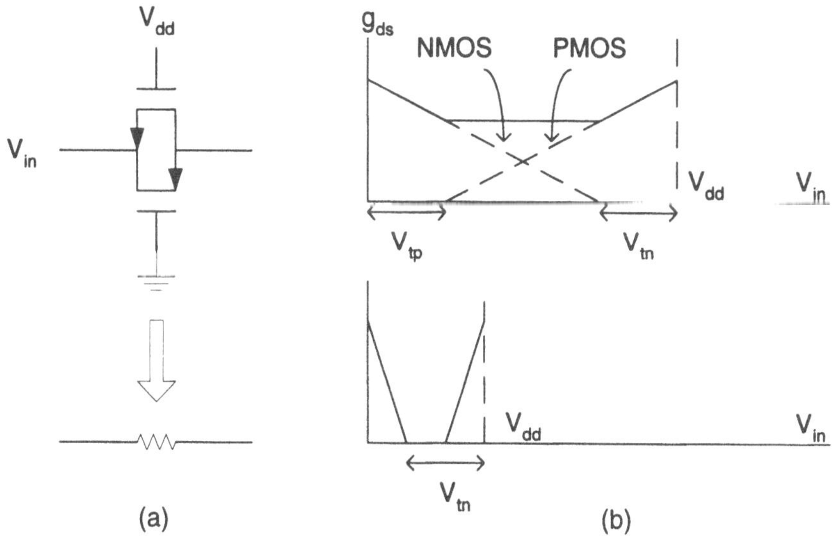

Figure 5.3. A Typical CMOS Switch and its Conductance

When $V_{dd} > V_{T,n} + V_{T,p}$, the total switch on-resistance is given by:

$$R_{on} = \frac{1}{\mu_n C_{ox}(\frac{W}{L})(V_{dd} - V_{in} - V_{T,n}) + \mu_p C_{ox}(\frac{W}{L})(V_{in} - V_{T,p})} \tag{5.7}$$

This resistance along with the sampling capacitance determines the acquisition time of the sampling circuit. For a fixed sampling capacitor, the only way to improve sampling speed is to reduce R_{on}. However, the values of $(V_{dd} - V_{in} - V_{T,n})$ and $(V_{in} - V_{T,p})$ are very limited at low supply voltages, and μ_n, μ_p and C_{ox} are fixed values at a given technology, so large (W/L) devices have to be used. Then some other problems are introduced. The larger depletion capacitance can couple more clock feedthrough noise, and charge injection may become dominant in error sources even if the fully differential topology is used.

Another error that appears in low supply voltage switches stems from the variation of the switch on-resistance with the input voltage. For high-frequency inputs, this variation introduces input-dependent phase shift and hence harmonic distortions [Razavi, 1995].

New circuit techniques are therefore required to allow the proper operation of analog switches at very low supply voltages while keeping similar performance in terms of speed, SNR, etc.

Three solutions to the switch driving problem at very low supply voltages have been proposed in the recent publications: the use of low threshold devices [Matsuya and Tamada, 1994, Adachi et al., 1990], the use of on-chip clock boosting [Dickson, 1976, Castello and Tomasini, 1991, Wu et al., 1996, Cho and Gray, 1995, Abo and Gray, 1999, Dessouky and Kaiser, 2001], and the switched-opamp technique [Crols and Steyaert, 1994, Baschirotto and Castello, 1997, Peluso et al., 1998, Baschirotto et al., 1994, Waltari and Halonen, 2001].

3. Low-V_T Devices

In order to get enough switch over-drive at very low supply voltages, low-V_T devices can be used as switches. This approach, however, has several disadvantages. The first one is the high cost of technologies with low-V_T devices. Special fabrication process is required for low-V_T devices. As mentioned before, all digital, analog, and memory circuits will use the same CMOS technology in a SoC design. This means the whole system has to be fabricated in more expensive technologies because of very limited number of analog switches, which is obviously not cost-effective.

The second problem associate with low-V_T devices is high unwanted leakage currents [Crols and Steyaert, 1994]. It may results in a deformation of the transfer characteristics in circuits.

Another problem in using special low-V_T devices comes from threshold voltage variation [Abo, 1999]. When threshold V_T is scaled down, the variation in device V_T becomes larger and more difficult to control due to the statistical variation of the number of dopant atoms in channel region. This causes matching problems in precision analog circuits.

4. Clock Boosting

Clock boosting is a circuit technique to generate a voltage higher than the power supply voltage by means of an on-chip voltage multiplier. The generated high voltage is used to drive critical analog switches. This solution is prevalent nowadays in low voltage designs. Earlier clock boosting circuits resulted in relative terminal voltage stress exceeding the supply by a large margin, thus long-term reliability became a problem. Recently, some new clock boosting circuits have been proposed [Abo, 1999, Dessouky and Kaiser, 2001] which avoid gate dielectric overstress.

Shown in Figure 5.4 is the locally bootstrapped switch circuit used to implement a 1-V $\Delta\sum$ modulator [Dessouky and Kaiser, 2001]. It functions as follows. The $N - switch$ is the main switch, and the other switches $S_1 - S_5$ and capacitor C are for bootstrapping purpose. During phase ϕ_2, C is charged to V_{dd} and $N - switch$ is off. During phase ϕ_1, the precharged capacitor C is connected between the gate and source of $N - switch$, therefore the V_{gs} of $N - switch$ keeps constant as V_{dd}, independent of the input signal V_{in}. In this way, a constant low on-resistance is established from the drain to source of the $N - switch$. And although the absolute voltage applied to the gate will exceed supply voltage for a positive input signal, none of the relative terminal-to-terminal device voltages exceeds V_{dd}, so it will not degrade the lifetime of the device [Abo, 1999].

However, in future deep submicron CMOS technologies, devices will not be able to withstand the bootstrapped voltage, so it will be imperative to avoid such clock boost circuits in very low voltage circuit design.

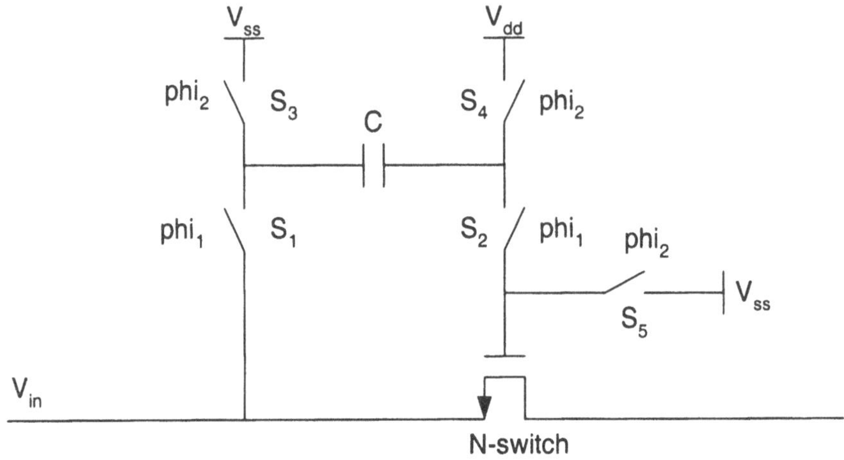

Figure 5.4. A Locally Bootstrapped Switch Circuit

Another disadvantage of this technique is the power consumption and area associated with the clock boost circuits, because each analog switch in the signal path is usually driven by a separate clock boost circuit.

5. Switched-Opamp

Recently, the switched-opamp technique has been developed to allow one to design switched-capacitor circuits without the use of clock boosting or low-V_T devices. Several circuits including filters [Crols and Steyaert, 1994, Baschirotto and Castello, 1997], $\Delta\sum$ modulators [Peluso et al., 1998], and ADCs [Waltari and Halonen, 2001] have been reported in open literature.

The switched-opamp technique was first introduced in [Crols and Steyaert, 1994]. It is based on the observation that in typical switched-capacitor circuits (as in Figure 5.5), the switches suffering from limited over-drive are usually found at the output of the opamps (such as S_5), while all other switches ($S_1 - S_4$) are connected to a fixed voltage (reference voltage or ground) and are thus not problematic. The core idea of switched-opamp technique is to replace those switches at the output of opamps with a switchable opamp, while leaving all other switches unchanged.

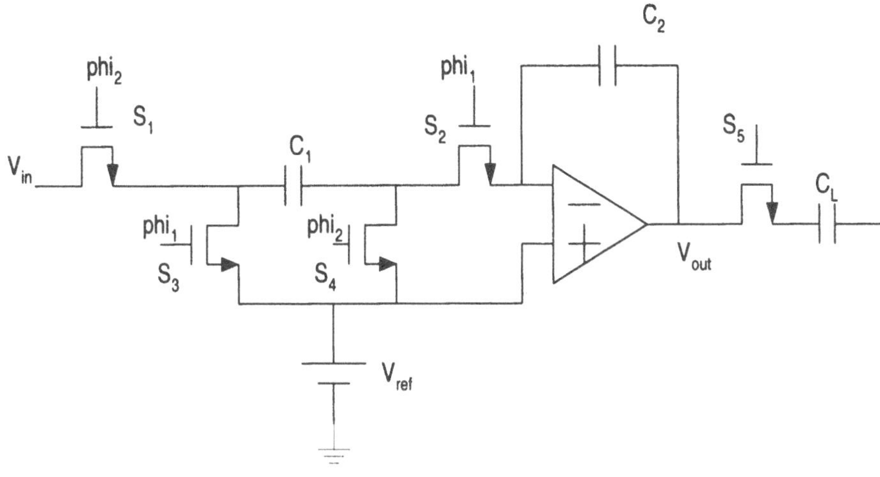

Figure 5.5. A Typical Switched-capacitor Integrator

Figure 5.6 shows the original switched-opamp integrator proposed in [Crols and Steyaert, 1994]. In this circuit, the load capacitor C_L is directly connected to the output of opamp which can be switched on and off during different phases. In this way, the switchable opamp acts as an output switch.

Notice that in this circuit, the input and output common mode levels of the opamp are equal. As a consequence, the output signal swing is limited. Another limitation is that the opamp is turned off completely during one clock phase which limits the maximum sampling frequency. In [Baschirotto and Castello, 1997], a modified switched-opamp technique is proposed which overcomes above limitations. In this circuit, the input and output common levels of the opamp are separated. At the opamp outputs, the dc level is taken in the middle of the full output swing, while at the opamp inputs, it is set as V_{ss}. The opamp also employs a two-stage topology where only the output stage is turned on and off. This increases the maximum sampling frequency to 1.8MHz. Another modification is that a fully differential topology is used in the circuit.

In contrast to low-V_T devices and clock boosting techniques, the switched-opamp technique is fully compatible with future deep submicron technologies. It has also power and area advantages over the clock boosting technique due to its switched-opamp nature and the elimina-

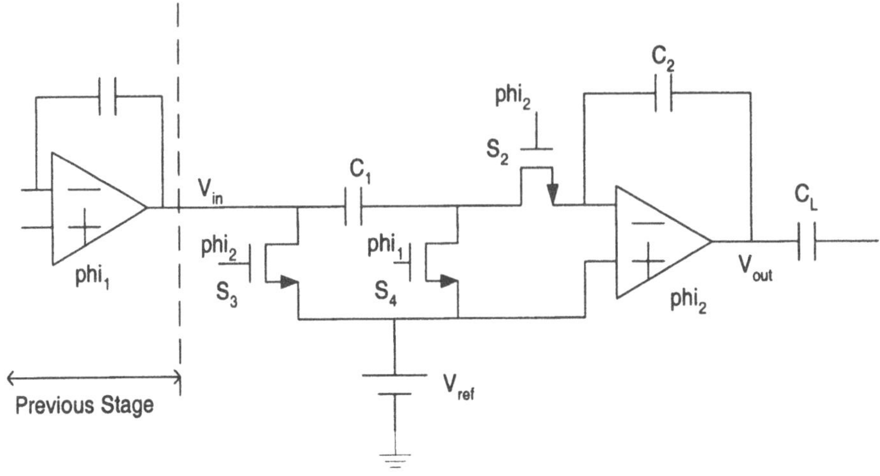

Figure 5.6. A Original Switched-Opamp Integrator

tion of bulky on-chip clock boost circuits. It appears to be the most attractive solution to very low voltage analog circuit designs. Therefore it is chosen as the method of implementation for our prototype pipeline ADC.

6. A Modified Switched-opamp Technique

One problem with the switched-opamp technique is the lack of a good input sampling switch, which can not be avoided because it is not preceded by another switched-opamp circuit. Another limitation of this technique stems from the need to switch off the entire or part of the opamp, which limits its maximum achievable operating frequency. In this section, some modifications to the conventional switched-opamp technique are proposed to solve above problems.

6.1 Input Stage

A major problem in switched-opamp circuits is the lack of a series switch to be connected to the input signal. Recently, a few solutions have been reported [Baschirotto et al., 1998, Waltari and Halonen, 2001], but they all have their own disadvantages.

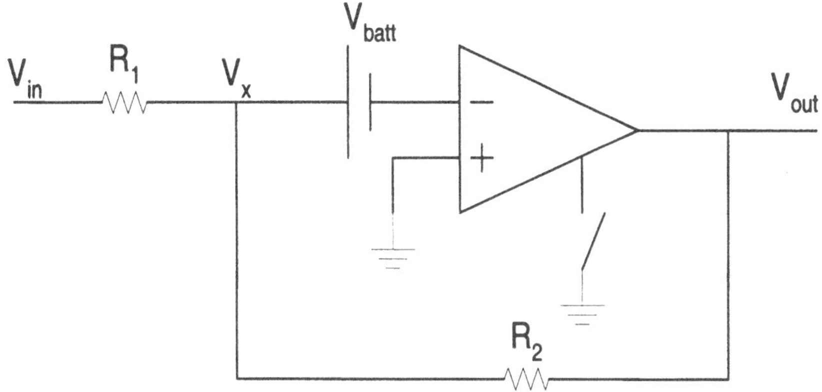

Figure 5.7. The Conceptual Scheme of the Active Series Switch

In [Baschirotto et al., 1998], an active series switch is proposed. As shown in Figure 5.7, it is basically a switched-opamp based inverting amplifier. The constant voltage source $V_{batt} = V_{dd}/2$, implemented as a switched-capacitor level shifter, is inserted in series with the opamp inverting input node. In this configuration, the node V_x acts as a virtual ground set to $V_{dd}/2$, so both the input and output common mode can be set as $V_{dd}/2$, guaranteeing rail-to-rail operation. Notice that no biasing current flows in resistor R_1 and R_2.

The problem with this solution is the need of an opamp driving resistive load, which tends to consume large power and limit the operating speed.

In [Waltari and Halonen, 2001], a passive input interface is proposed as shown in Figure 5.8. It consists of switches S_1, S_2, S_3, the resistor R and the capacitor C. The basic idea is to attenuate the input signal by the ratio of C_s and C during sampling. And since the input of pipeline ADC is purely capacitive, using a decoupling capacitor causes only the loss of DC level.

Several problems exist in this solution. The major one is the attenuation of the input signal, which causes the degradation of SNR. Another problem is the signal feedthrough from V_{in} in the hold phase. The parasitic capacitance of switches S_1 and S_2 also produces harmonic distortions during the sampling.

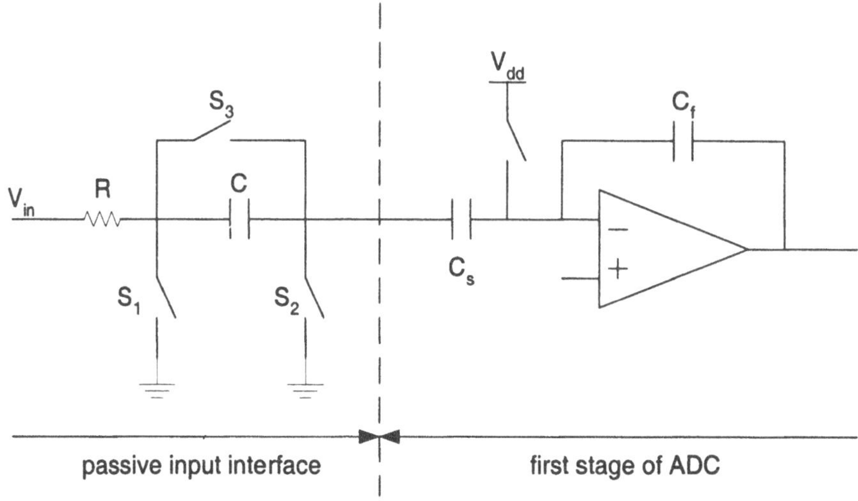

Figure 5.8. A Passive Input Interface with Part of the Pipeline Stage

In this section, a novel input stage is proposed to solve the input interface problem in switched-opamp circuits, without sacrificing the performance in speed, SNR, and power consumption. As shown in Figure 5.9, it combines a S/H and a gain amplifier. The core idea of this implementation is to limit the input signal range and amplify it by the front S/H stage of the pipeline ADC. It is a reasonable configuration in SoC design since typically the ADC in such systems is preceded by a variable gain amplifier (VGA), thus it is easy to migrate some gain from the VGA to the ADC.

This circuit works as follows. During sampling phase ϕ_1, the input signal charges the sampling capacitors labeled C_S, while the opamp is turned off and its output is shorted to V_{ss}. During holding phase ϕ_2, the opamp turns on, and the left plate of sampling capacitors is connected to V_{ss}, so the charge on C_S is transferred to the feedback capacitors C_F. The gain is set by the ratio of C_S/C_F. By properly choosing this gain, the input signal can be in a reasonable range, then the input switches can have enough over-drive voltage.

The capacitor $C1$ and other switches in this circuit are used for level shifting. The reason for adding this part is because there exists a common mode level change between the input signal and the input nodes of the

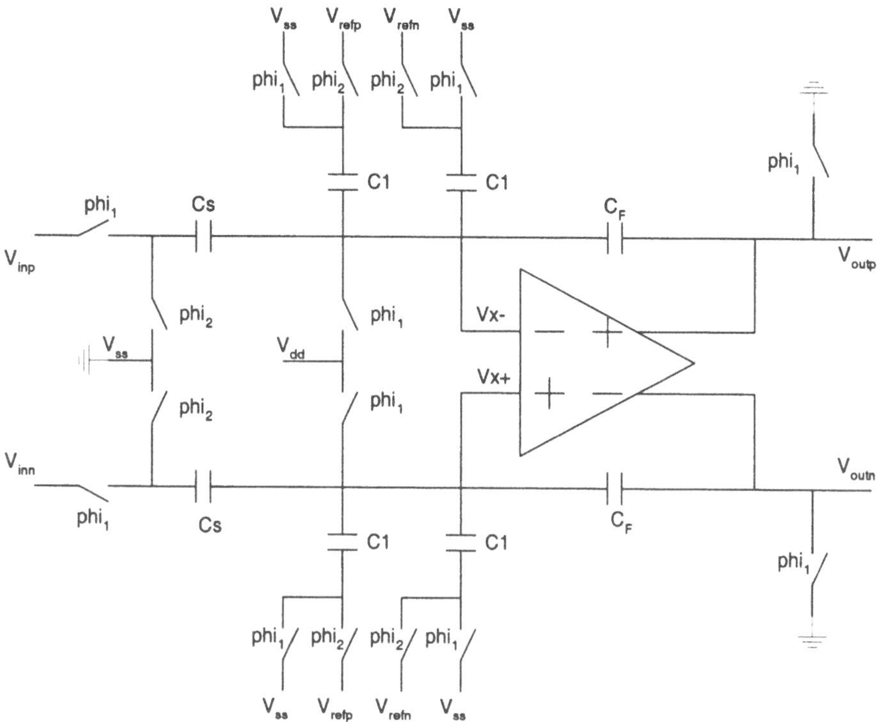

Figure 5.9. The Proposed Switched-opamp S/H with Gain Stage

opamp. In low voltage circuits, the opamp input common mode voltage is usually close to the negative or positive rail. In this circuit, it is set at positive supply rail V_{dd} as will be explained in the next section.

The principle of the level shifting will be clear from the charge equations. If we assume $C_F = C$, $C_S = 2C$, $C1 = C/2$, the common mode level of the input signal is $V_{cm,in} = (V_{inp} + V_{inn})/2$, and the output common mode level is $V_{cm,out} = (V_{outp} + V_{outn})/2$, and further assume $V_{cm,in} = V_{cm,out} = (V_{refp} + V_{refn})/2$, then based on charge conservation principle we can write:

$$2C * (V_{dd} - V_{inp}) + 2 * C/2 * (V_{dd} - V_{ss}) + C * (V_{dd} - V_{ss}) = 2C * (V_{x-} - V_{ss}) + C/2 * (V_{x-} - V_{refp}) + C/2 * (V_{x-} - V_{refn}) + C * (V_{x-} - V_{outp})$$

and

$$2C * (V_{dd} - V_{inn}) + 2 * C/2 * (V_{dd} - V_{ss}) + C * (V_{dd} - V_{ss}) = 2C * (V_{x+} - V_{ss}) + C/2 * (V_{x+} - V_{refp}) + C/2 * (V_{x+} - V_{refn}) + C * (V_{x+} - V_{outn})$$

They can be simplified as

$$4V_{dd} - 2V_{inp} = 4V_{x-} - V_{cm} - V_{outp} \tag{5.8}$$

$$4V_{dd} - 2V_{inn} = 4V_{x+} - V_{cm} - V_{outn} \tag{5.9}$$

From (5.9)-(5.8), since $V_{x-} = V_{x+}$ because of the virtual ground principle of opamp, we can get:

$$V_{out} = V_{outp} - V_{outn} = 2 * (V_{inp} - V_{inn}) = 2 * V_{in} \tag{5.10}$$

which means a gain-2-amplifier.

From (5.9)+(5.8) and using $V_{x-} = V_{x+}$, we get $V_x = V_{dd}$, which proves that the input common mode of opamp is truly set at the positive supply rail.

This novel input stage configuration has several advantages. In contrast to the passive input switch approach, it does not degrade the SNR of the input signal. Compared to the active series switch approach, this input interface offers a better linearity, and is more power efficient.

6.2 High Speed Design Techniques

In the original switched-opamp technique, the opamp is fully turned off during the idle phase and needs a very long turn-on time during the other phase, consequently its working frequency is very low. In the modified version, only the output stage of the opamp is turned off, so the maximum sampling frequency is increased. However, the clock frequencies of these circuits are still well below video rates and the switched-opamp technique is currently labeled as "low-speed". In this section, the speed limitation of conventional switched-opamp techniques is analyzed, and several modifications are proposed to improve the speed.

In low supply voltage circuits, the two-stage opamp topology is usually preferred, since the conventional single stage topologies such as folded-cascode or telescopic cascode opamp can not provide both high gain and large output signal swing at the same time. Moreover, the two stage topology offers higher switching speed because only the output stage of opamp needs to be turned off. Shown in Figure 5.10(a) is a typical output stage (single ended version) used in switchable opamps. During phase ϕ_1, S_1 is on while S_2 is off, the output stage is turned on and the load capacitor C_L is charged to the desired output voltage V_{out}. During phase ϕ_2, S_1 is off, shutting off the output stage, while S_2 is on, pulling the output to V_{ss}.

It is obviously that the output stage is the major limiting factor of sampling frequencies in the switched-opamp circuits. Figure 5.10(b) shows the equivalent circuit of the output stage. In this circuit, the load device PMOS device M_p is replaced with a resistor R, and the NMOS device is ignored during the short period after the switch S_1 is just turned on. In this short period, V_{dd} charges the load capacitor C_L through the resistor R, therefore it is the time constant of $\tau = R \cdot C_L$ that determines the turn-on time of the opamp. As an example, if $R = 10K\Omega$, $C_L = 0.5pF$, then $\tau = 5ns$. For the output signal settling to 8bit accuracy, $t = 6\tau = 30ns$ is required, which limits the maximum sampling rate to 16MHz.

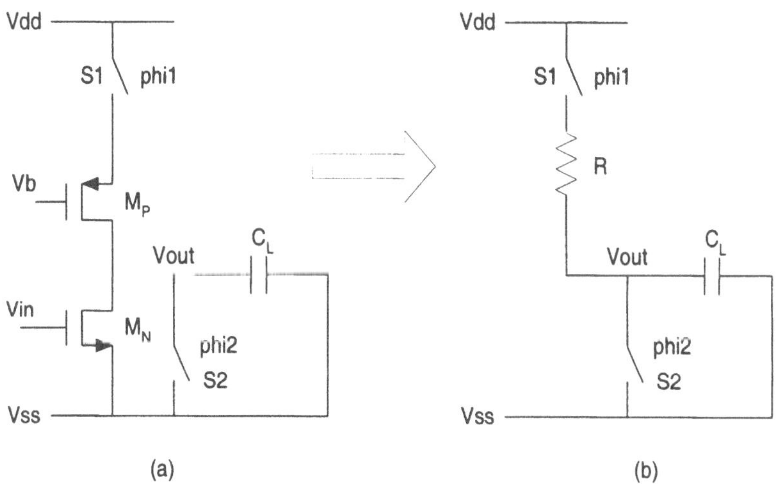

Figure 5.10. The Switchable Output Stage of Opamp

To increase the sampling rate of the switchable output stage, one needs to reduce the value of R, C_L or both. Usually in the circuit, the minimum C_L value is set by the thermal noise constraint. The only freedom in the design is the drain to source resistance R of the load device.

Two approaches are proposed to reduce R in the output stage. First, the channel length of load devices are chosen to be minimum. Although this is usually avoided in conventional opamp design for DC gain reason, it is quite useful in increasing the speed of switched-opamp circuits. The loss of DC gain because of the smaller output impedance in the output stage can be compensated by the folded-cascode first stage as shown in Figure 5.11, along with the increasing of the biasing current in the output stage. The switch in output stage is also made wide to reduce the switch on-resistance which is in series with R.

The second approach is to add another pair of load devices to the output stage (M_{7c} and M_{7d} in Figure 5.11). These devices are turned on during phase ϕ_3 which is only the beginning part of ϕ_1. They increase the biasing current of the output stage temporarily to further reduce the opamp turn-on time. The power consumption of these added devices is small since they are only used for a very short duration in each clock period. The final configuration of the improved switchable opamp is shown in Figure 5.11.

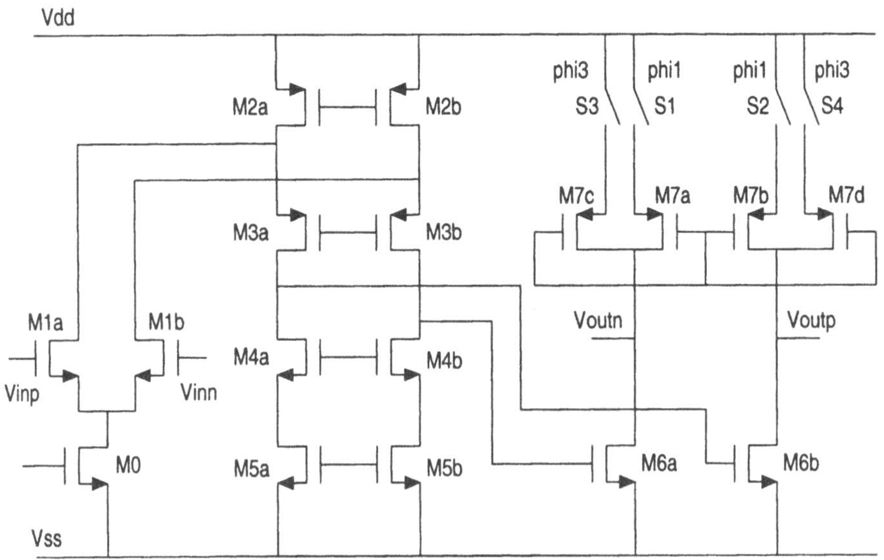

Figure 5.11. The Two-stage Opamp Used in SO Circuits

7. Summary

In this chapter, the need for very low voltage analog circuit design is explained. Several design challenges are discussed, and the switched-opamp technique seems to be the most attractive solution to low voltage analog circuit designs. Several modifications to the conventional switched-opamp technique have been proposed to improve the operating speed, and to bring the input signal to the circuit.

Chapter 6

ADC FOR BLUETOOTH/WLAN(FHSS)/HOMERF

As a demonstration of the improved switched-opamp technique described in the previous chapter, an 8-bit, 20MS/s, 1.5V pipeline ADC was designed and implemented in a $0.18\mu m$ CMOS technology. Design issues of the ADC in both system and circuit levels are discussed in this chapter.

1. Applications

The prototype ADC is targeted for low power low cost wireless applications such as Bluetooth/WLAN(FHSS)/HomeRF. As shown in Figure 6.1, these three standards are similar to each other. They are all short range radio technologies for connections between mobile PCs, cellular phones and other portable devices. They also share the same receiver architecture. Therefore, one ADC can be designed for all three applications.

Table 6.1. Summary of Bluetooth, WLAN (FHSS), and HomeRF Standards

Standard	Multiple Access	Frequency Range	Channel Spacing	Modulation Scheme	Data Rate
WLAN(FHSS)	FHSS	2.4GHz	1MHz	GFSK	1, 2Mbps
Bluetooth	FHSS	2.4GHz	1MHz	GFSK	1Mbps
HomeRF	FHSS	2.4GHz	5MHz	GFSK	5, 10Mbps

As an example, the Bluetooth standard is briefly introduced, and the ADC requirement in a Bluetooth receiver is discussed. Bluetooth is a new universal radio interface enabling electronic devices from different manufacturers to connect and communicate wirelessly via short-range connections [Haartsen and Mattisson, 2000]. It was originally proposed by a special interest group (SIG) formed by five companies (Intel, IBM, Ericsson, Nokia, and Toshiba) in February 1998, and named after a Danish king Harald Blatand from the tenth century who united Denmark and Norway. Currently there are several thousands companies in this SIG, some of which have recently introduced their demo products.

Bluetooth enabled devices support *ad hoc* connectivity with no master device required to provide call setup and networking functions. It operates at unlicensed frequency bands (usually the Industrial-Scientific-Medical (ISM) band from 2.4GHz to 2.48GHz). Bluetooth radio uses frequency hopping (FH) spread spectrum, in which each channel has 1MHz bandwidth. Bluetooth channels use a FH/time division duplex (FH/TDD) scheme. If Gaussian-shaped frequency shift keying (GFSK) modulation is used, then a symbol rate of 1 Mb/s can be achieved.

Figure 6.1 shows a receiver architecture used in Bluetooth. Since Bluetooth is mainly used in portable and low cost devices, its radio transceiver must be small and operate at low voltages with low power consumptions. It is preferable to take a single chip solution and be implemented in advanced deep submicron technologies. The switched-opamp technique is suitable for designing ADCs in Bluetooth receivers.

The prototype ADC has a 8-bit resolution, and the sampling frequency is designed to be over 10MHz, which meets the requirements of all three standards (Bluetooth/WLAN (FHSS)/HomeRF).

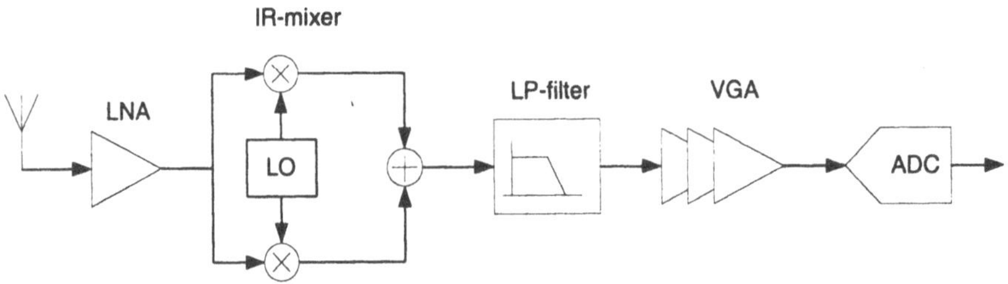

Figure 6.1. A Bluetooth Receiver

2. System Level Design

The conventional 1.5b/stage pipeline architecture discussed in Chapter 3, Section 4.2 is used in this design. As shown in Figure 6.2, the ADC consists of an input S/H plus gain stage, six 1.5-bit pipeline stages, and a final 2-bit flash stage. Each pipeline stage and the flash stage resolve two bits with a sub-ADC, and the final 8-bit digital outputs are obtained from these 14-bit codes through a digital correction circuit. As discussed before, the 1.5bit/stage architecture reduces the total power consumption of ADC by relaxing the accuracy requirement of comparators.

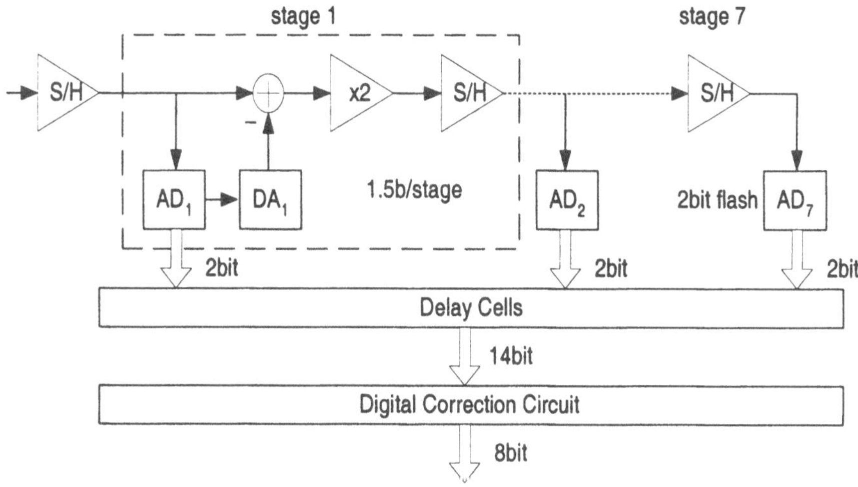

Figure 6.2. Block Diagram of the 8-bit Pipeline ADC

To achieve the desired effective number of bit (ENOB), the non-ideal errors in each stage of ADC must be in a certain range. Ideally, the transfer function of the MDAC in this architecture can be written as:

$$V_{out} = 2 \cdot V_{in} \pm V_{ref} \tag{6.1}$$

where V_{in} and V_{out} are the input and output signals, respectively, and V_{ref} is the reference voltage.

After adding non-ideal effects such as capacitor mismatch, finite opamp gain and settling, and thermal noise, it becomes [Abo, 1999]

$$V_{out} = (1 + \frac{\Delta G}{G} + \frac{1}{2}\frac{\Delta C}{C})2V_{in} \pm (1 + \frac{\Delta G}{G} + \frac{\Delta C}{C})V_{ref} + v_n \tag{6.2}$$

where $\Delta G/G$ is the combined gain error caused by finite opamp gain and settling, $\Delta C/C$ is the mismatch between sampling capacitors C_s and feedback capacitors C_f, v_n is the output referred thermal noise.

For the first stage, the total error caused by all non-ideal effects can not exceed 0.5LSB at the $(N-1)$-bit level (7bit in our case) from which the requirements on the opamp can be derived. The requirement on the following stages is more relaxed, since the noise and non-ideal effects is divided by the inter-stage gain of previous stages when referred to the input.

The required specifications on the opamp used in the first pipeline stage are listed in Table 6.2.

Table 6.2. Required Specifications of the First Opamp

Open loop DC gain	$\geq 60dB$
Unit gain frequency	$\geq 200MHz$
Phase margin	$\geq 70^o$
Slew rate	$\geq 200V/\mu S$
Load capacitance	$\geq 1pF$
Differential output swing	$\geq 1V_{pp}$
Supply Voltage	$\leq 1.5V$

3. A Switched-opamp MDAC

The switched-opamp implementation of the MDAC in the 1.5b/stage pipeline ADC is shown in Figure 6.3.

In this circuit, the opamp input common mode voltage is set at V_{dd}, which is a better choice than V_{ss}. If it is set at V_{ss}, the voltage spikes during transients cause the opamp input nodes to go beneath V_{ss}, resulting charge loss on capacitors. This is because in typical CMOS technologies, the p-substrate is tied to V_{ss}, and forms a pn junction with the drain/source of the NMOS switches at the opamp inputs. This diode could be forward biased by large negative spikes caused by the charge injection and the finite speed of the opamp [Baschirotto and Castello, 1997]. Therefore, V_{dd} is used as virtual ground in the opamp to avoid this problem.

The MDAC operates at a two-phase non-overlapping clock. During phase ϕ_1, opamp inputs are shorted to V_{dd}, while the outputs are pulled to V_{ss}, and the input signal is applied to the sampling capacitors $C_s(=2C)$ and the sub-ADC which is comprised of two comparators with switching thresholds set at $+V_{ref}/4$ and $-V_{ref}/4$, respectively (the sub-ADC is not shown in Figure 6.3). At the end of ϕ_{1a}, the signal is sampled and the comparators in the sub-ADC are latched. During phase ϕ_2, the bottom plate of C_s is connected to V_{ss}, transferring its charge to the feedback capacitor C_f. At the same time, the capacitor C_1 is connected to $(V_{refp}-V_{refn}) = +V_{ref}$, $(V_{refn}-V_{refp}) = -V_{ref}$, or $(V_{refn}-V_{refn}) = 0$ based on the value of X, Y, and Z which are generated by the sub-ADC as follows:

$$subADCoutput = \begin{cases} X & \text{when } V_{in} > V_{ref}/4 \\ Y & \text{when } V_{in} < -V_{ref}/4 \\ Z & \text{when } -V_{ref}/4 < V_{in} < V_{ref}/4 \end{cases}$$

As a result, the residue of the MDAC is given by:

$$V_{out} = \begin{cases} C_s/C_f \cdot V_{in} - C_1/C_f \cdot V_{ref} & whenV_{in} > V_{ref}/4 \\ C_s/C_f \cdot V_{in} + C_1/C_f \cdot V_{ref} & whenV_{in} < -V_{ref}/4 \\ C_s/C_f \cdot V_{in} & when - V_{ref}/4 < V_{in} < V_{ref}/4 \end{cases}$$

4. Opamp Design

A two-stage topology is preferred in low voltage opamp design since it offers large open loop DC gain and rail-to-rail output swing. It has also an intrinsic advantage in switched-opamp circuits that it separates the input stage from the switchable output stage, resulting higher operating speed.

Shown in Figure 6.4 is the opamp employed in this design. Since the virtual ground of the opamp is set as V_{dd}, NMOS input differential pair is used. As discussed before, to improve switching speed, the DC gain of the output stage has to be low, therefore a folded-cascode first stage is employed. The output stage is made switchable with two switches S_1 and S_2.

Cascode compensation is used in the opamp, since it offers inherently higher bandwidth than the traditional Miller compensation. The disadvantage is that it creates a three-hole system, which conflicts the design

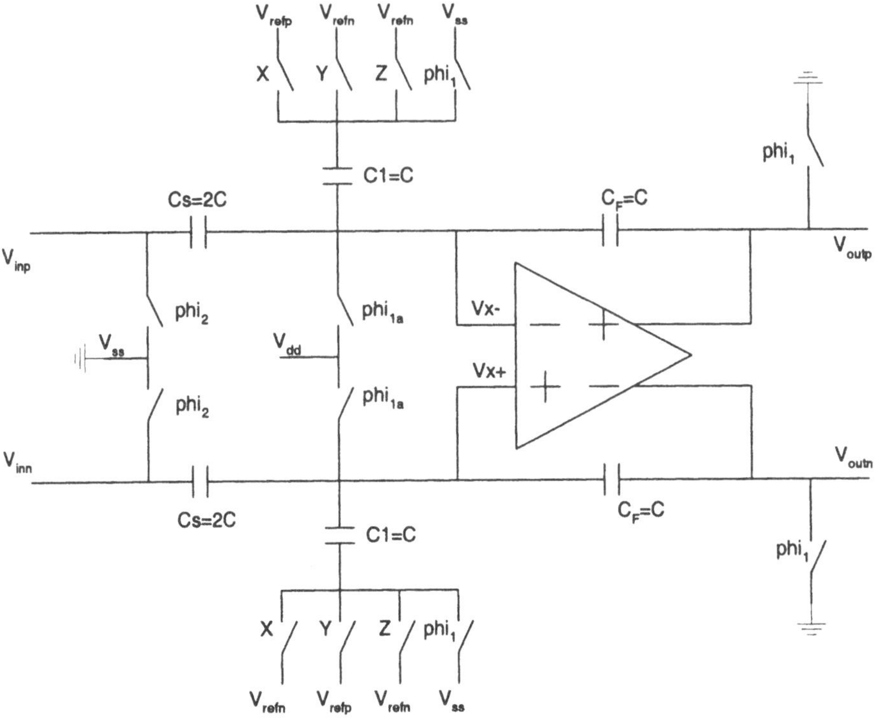

Figure 6.3. Block Diagram of the Switched-opamp MDAC

and makes it difficult to find a direct relationship between phase margin and transient step-response.

4.1 DC Gain

The open loop DC gain of the opamp can be written as:

$$A_{DC,1} = g_{m1} \cdot [g_{m4} \cdot r_{o4} \cdot r_{o5}]//[g_{m3} \cdot r_{o3} \cdot (r_{o1}//r_{o2})]$$

$$= \frac{(g_{m1}r_{o1})(g_{m3}r_{o3})(g_{m4}r_{o4})r_{o2}r_{o5}}{(g_{m3}r_{o3})r_{o1}r_{o2} + (g_{m4}r_{o4})r_{o1}r_{o5} + (g_{m4}r_{o4})r_{o2}r_{o5}} \tag{6.3}$$

$$A_{DC,2} = g_{m6} \cdot (r_{o6}//r_{o7}) = g_{m6} \cdot \frac{r_{o6} \cdot r_{o7}}{r_{o6} + r_{o7}} \tag{6.4}$$

$$A_{DC,T} = A_{DC,1} \cdot A_{DC,2} \tag{6.5}$$

where g_m and r_o are the device's transconductance and output resistance, respectively. Since the transistor output resistance is proportional to the channel length, the load devices M_2 and M_5 can have long channels to maximum the DC gain of the first stage, without sacrificing the bandwidth.

4.2 Frequency Response

From small-signal analysis, the closed loop transfer function of the opamp can be derived as [Abo, 1999]:

$$A(s) = \frac{1}{f} \frac{1 - s^2 \frac{C_c C_3}{g_{m3} g_{m6}}}{1 + s\frac{C_c}{f g_{m1}} + s^2 \frac{C_3}{g_{m6}}\left(\frac{C_c + C_6}{f g_{m1}} - \frac{C_c}{g_{m3}}\right) + s^3 \frac{C_3(C_c C_1 + C_c C_6 + C_1 C_6)}{f g_{m1} g_{m3} g_{m6}}} \tag{6.6}$$

where C_c is the compensation capacitance, C_1, C_3 and C_6 are the sum of parasitic and load capacitance at the drain of devices M_1, M_3 and M_6, respectively, and f is the feedback factor. As can be seen from the equation, this is a three-pole system. Proper compensation is required to make the close loop system have critically damped step response.

The settling speed of an opamp is determined by its gain-bandwidth product (GBW). The GBW of the switched-opamp MDAC is given by

$$GBW_{MDAC} = \frac{g_{m,eff} \cdot f}{2\pi C_{L,eff}} \tag{6.7}$$

with

$$f = \frac{C_F}{C_F + C_s + C_{in} + C_1} \tag{6.8}$$

and

$$C_{L,eff} = C_L + C_F \cdot (C_s + C_1 + C_{in})/(C_f + C_s + C_1 + C_{in}) \tag{6.9}$$

in which C_L is the total load capacitance at the opamp output, C_{in} is the opamp input capacitance. Substituting (6.8) and (6.9) into (6.7), the closed-loop bandwidth can be written as

$$GBW_{MDAC} = \frac{g_{m,eff}}{2\pi[C_s + C_1 + C_{in} + C_L \cdot (C_f + C_s + C_1 + C_{in})/C_f]} \tag{6.10}$$

In the switched-opamp MDAC as shown in Figure 6.3, the feedback factor is less than $1/4$, which makes the switched-opamp technique inherently slower than its switched-capacitor counterpart. However, it also makes it relatively easier to achieve enough phase margin in the opamp design.

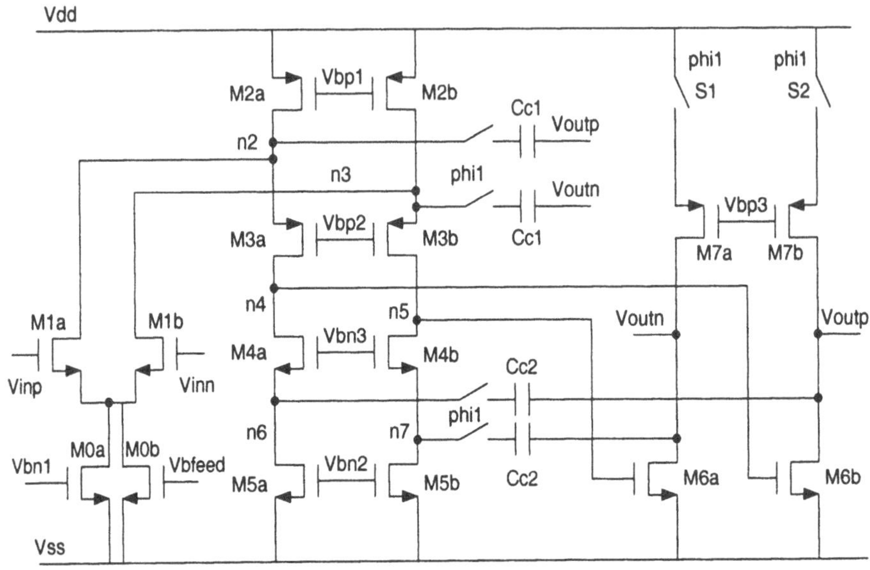

Figure 6.4. The Designed Two-stage Switchable Opamp

4.3 Slew Rate

The slew rate is the maximum rate at which the output changes when a large input signal is applied to the opamp [Jones and Martin, 1996]. Slew happens when opamp outputs are unable to track the large rate of changes in the inputs. There are two different slew rates in a two-stage opamp. For the folded-cascode first stage, it is determined by the tail current of M_1 and M_2 and the compensation capacitor:

$$SR_1 = I_{M0}/C_c \tag{6.11}$$

For the common-source output stage, it is given by:

$$SR_2 = \frac{I_{M7}}{C_c + C_L} \tag{6.12}$$

The slew rate of the opamp is determined by the lesser of these two values: $SR = min(SR_1, SR_2)$. To increase SR, the biasing current of both stages needs to be increased. In the switched-opamp MDAC, the opamp outputs need to jump from V_{ss} to the final value in every amplifying phase, so the opamp needs more current than its switched-capacitor counterpart. However, in switched-opamp circuits the output stage is turned off during the idle phase, so the average power consumption is usually still lower than that of switched-capacitor circuits.

4.4 Noise

The input-referred thermal noise of a MOS transistor in the active region is modeled as:

$$V_{in,thermal}^2 = 4kT(\frac{2}{3})\frac{1}{g_m}\Delta f \tag{6.13}$$

where k is Boltzmann's constant ($1.38 \times 10^{-23} JK^{-1}$), T is the temperature in Kelvins, g_m is the transconductance of the device, and Δf is the frequency range. In the opamp as shown in Figure 6.4, the noise contribution of cascode devices M_3, M_4 and output devices M_6, M_7 is negligible. The input referred thermal noise power spectrum of the opamp can then be calculated as:

$$v_{in,thermal}^2 = 2 \cdot 4kT \cdot \frac{2}{3} \cdot (\frac{1}{g_{m1}} + \frac{g_{m2}}{g_{m1}^2} + \frac{g_{m5}}{g_{m1}^2}) \cdot \Delta f \tag{6.14}$$

The output referred noise can be obtained by multiplying the $V_{in,thermal}^2$ with the close loop gain of opamp and integrating it with frequency from DC to infinity:

$$V_{out,thermal}^2 = \int_0^\infty \frac{1}{f_F} \cdot A(f) \cdot 2 \cdot 4kT \cdot \frac{2}{3} \cdot (\frac{1}{g_{m1}} + \frac{g_{m2}}{g_{m1}^2} + \frac{g_{m5}}{g_{m1}^2}) \cdot df \tag{6.15}$$

where $A(f)$ is the same as in Equation 6.6, f_F is the feedback factor in the close loop system as given by:

$$f_F = \frac{C_f}{C_s + C_f + C_1 + C_{in}} \tag{6.16}$$

where C_{in} is the opamp input capacitance, and all other capacitors are shown in Figure 6.3.

Usually the feedback factor is fixed once the circuit topology is given. To decrease the total thermal noise, one can choose to increase g_{m1} or decrease g_{m2} and g_{m5}.

Flicker noise usually arises due to the traps of carriers in the semiconductor devices [Jones and Martin, 1996]. It is also referred as $1/f$ noise since it has a $1/f^{\alpha}$ spectral power density. The flicker noise of MOS devices can be modeled as

$$v_{flicker}^2(f) = \frac{K}{WLC_{ox}f} \tag{6.17}$$

The input referred flicker noise of the two-stage opamp can be calculated as

$$\begin{aligned} v_{in,flicker}^2(f) &= 2 \cdot \frac{K_n}{(WL)_1 C_{ox} f} + 2 \cdot \frac{K_p}{(WL)_2 C_{ox} f} (\frac{g_{m2}}{g_{m1}})^2 \\ &\quad +2 \cdot \frac{K_n}{(WL)_5 C_{ox} f} (\frac{g_{m5}}{g_{m1}})^2 \end{aligned}$$

where K_n and K_p are the flicker noise coefficients of NMOS and PMOS devices, respectively. To reduce flicker noise, the (WL) size of the transistors M_1, M_2, and M_5 can be made large.

4.5 Common-Mode Feedback (CMFB)

CMFB is required in fully-differential opamps to define the common-mode level at the output nodes. The conventional switched-capacitor CMFB is not suitable for low voltage applications since it requires signal-passing switches at the output nodes of opamp.

A suitable CMFB circuit used in this design is shown in Figure 6.5. It consists of a capacitive voltage divider, a switched-capacitor level-shifter, and an error amplifier. The divider, consisting of capacitors C_1 and C_2, samples the output common-mode voltage and applies it to the input of the error amplifier. Note that the divider is always connected to the opamp outputs, eliminating the need of signal-passing switches.

The capacitor C_3 and three switches S_1, S_2, and S_3 act as a level-shifter. In the proposed switched-opamp MDAC, the outputs of the opamp are pulled to V_{ss} during ϕ_1, while during ϕ_2 the output common-mode level is chosen to be in the middle of rails. This level-shifter is added to compensate the common-mode voltage change between two phases.

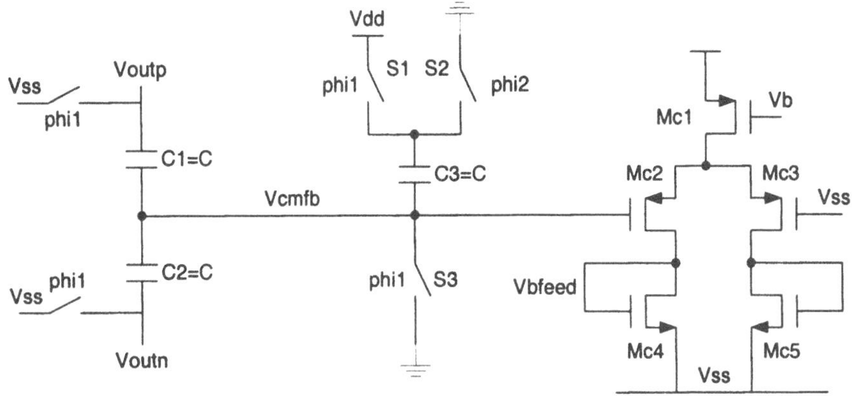

Figure 6.5. The Common-Mode Feedback Circuit

The error amplifier in the proposed CMFB consists of a simple differential pair with two diode-connected load transistors. The sensed output common-mode voltage is first level-shifted by $V_{dd}/2$, then compared to the desired level (V_{ss}). The amplified error signal V_{bfeed} is applied to the first stage of the opamp. It completes a negative feedback loop.

5. Comparator Design

As discussed in Chapter 3, because of digital correction mechanism employed in the 1.5b/stage pipeline architecture, low accuracy dynamic comparators can be used to reduce the power consumption.

The comparator used in this design is shown in Figure 6.6.

It has a threshold voltage of $+V_{ref}/4$. By switching the connection of reference voltages, the other comparator with $-V_{ref}/4$ threshold can be realized in the same architecture. The reference voltages V_{refp1} and V_{refn1} are generated from a resistor string as shown in Figure 6.7. It is shared by all comparators in the 1.5-bit pipeline stages as well as the last 2-bit flash stage.

The comparator operates on a two phase clock. During phase ϕ_2, the reference voltages are sampled to the capacitors C_1 and C_2, while at the same time the input nodes of the dynamic latch are shorted and the common-mode level is set to V_{dd}. The input signal is pulled to V_{ss} by the previous switched-opamp stage. During phase ϕ_1, the bottom plates of

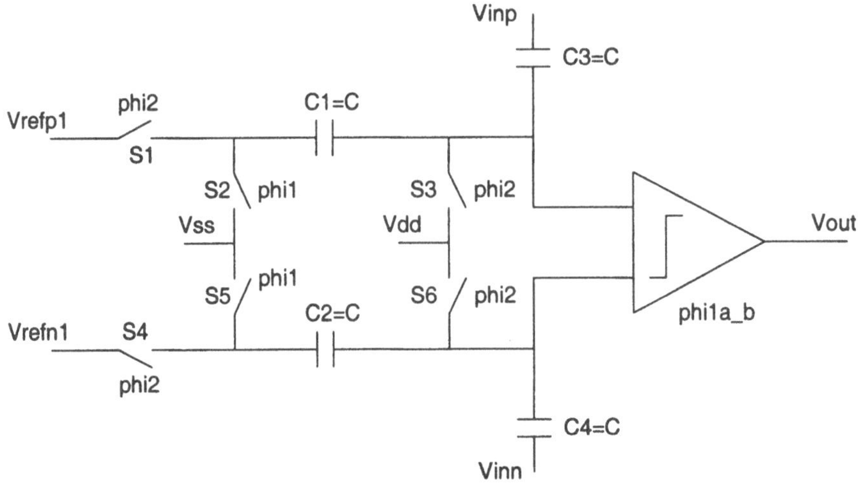

Figure 6.6. The Proposed Comparator

capacitors C_1 and C_2 are shorted to V_{ss}, and the input signal is applied to the latch through capacitors C_3 and C_4. Therefore a voltage level corresponding to $(V_{in} - V_{ref}/4)$ is sensed at the input of the dynamic latch. At the end of ϕ_{1a}, the comparator is latched to make a comparison.

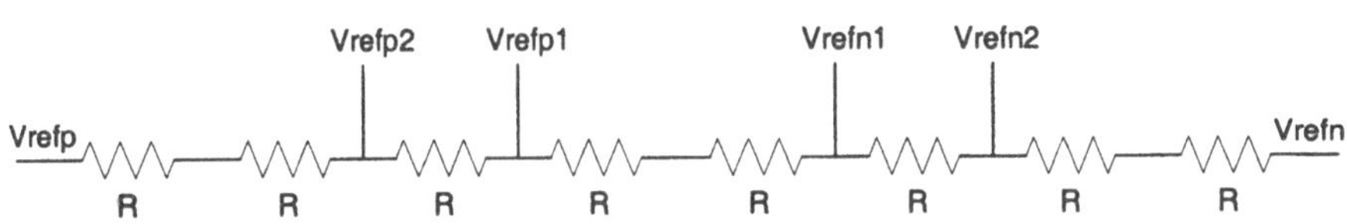

Figure 6.7. The Reference Voltage Generator

The dynamic latch used in the comparator is shown in Figure 6.8. It operates as follows. When ϕ_1 is low, the latch is in reset state, the tail current is off and the outputs are pulled to V_{dd}. When ϕ_1 goes high, the active phase starts. The tail current is on and the input differential pair functions. The output nodes settle to correct values from the metastable point.

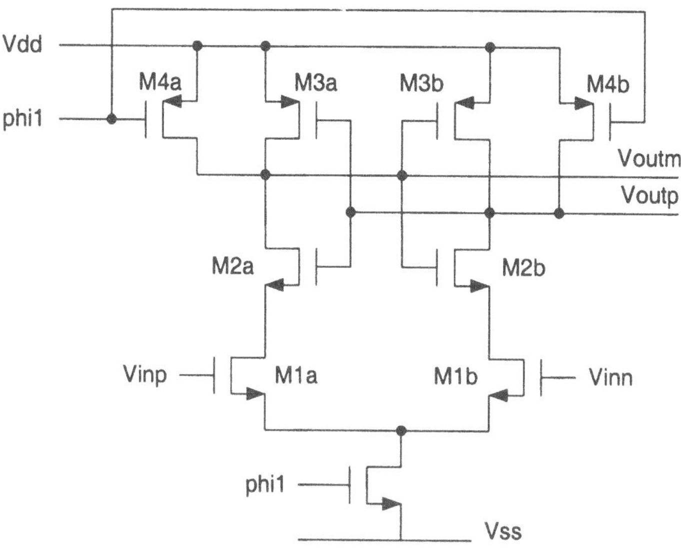

Figure 6.8. The Dynamic Latch Used in the Comparator

6. Clock Generator

The pipeline ADC operates with a two-phase, non-overlapping clock in order to reduce the signal-dependent charge injection errors. The clock timing diagram is illustrated in Figure 6.9.

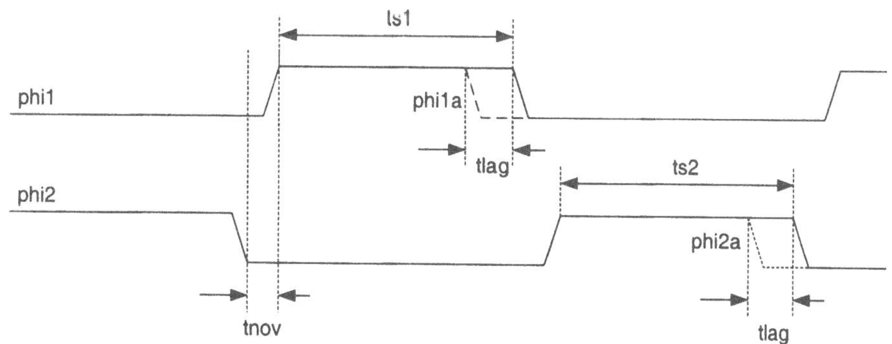

Figure 6.9. The Clock Waveform

Figure 6.10 shows the clock generator used in this design. The non-overlapping time t_{nov} and delay time t_{lag} in Figure 6.9 are controlled by the propagation delays in inverters and NAND gates. To keep the duration of t_{s1} and t_{s2} as equal as possible, the delay of inverter u_1 needs to be small.

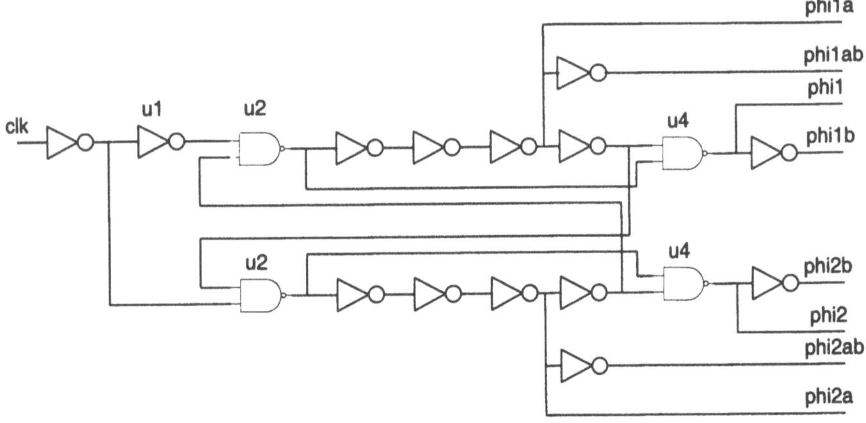

Figure 6.10. The Clock Generator

7. Digital Correction Circuit

Digital correction circuit is used to get the final 8-bit digital outputs from the 14-bit digital signals generated from 7 sub-ADC stages. It consists of a certain number of D-flip-flop delay cells and several 1-bit full adders.

8. Performance

The prototype switched-opamp pipeline ADC is designed and implemented with UMC $0.18\mu m$ mixed-mode/RF CMOS process. This process has the same typical threshold voltage of 0.5V for NMOS and PMOS devices. The maximum supply voltage is 1.8V. The designed 8-bit ADC operates at the typical voltage of 1.5V.

The chip layout is shown in Figure 6.11. The active die area is about $0.5\mu m \times 0.65\mu m$.

The ADC has been extensively simulated at both block and system levels, using HSPICE level 49 models. Shown in Figure 6.12 is the frequency response of the opamp used in the ADC. As can be seen, the DC gain is $78dB$, and the unit gain frequency is 258 MHz. Since in

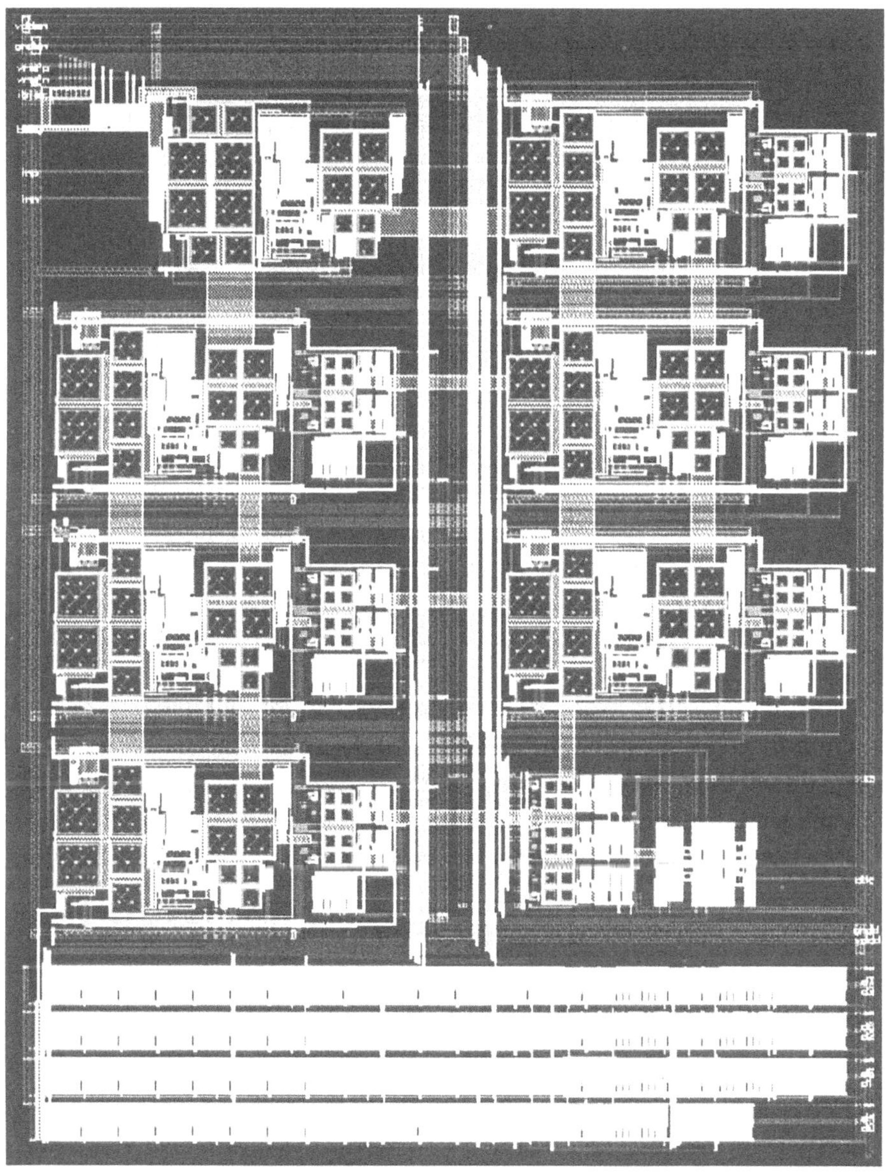

Figure 6.11. The Layout of the Implemented Pipeline ADC

the proposed switched-opamp MDAC, the feedback factor $f = 1/4$, we are only interested in the phase margin of the opamp when its gain is

Table 6.3. Performance Summary of the Switched-opamp Pipeline ADC

Parameter	Values
Technology	$0.18\mu m$ CMOS
Resolution	8bits
Conversion Rate	20 MS/s
SNDR	45 dB
Active Area	0.5mm x 0.65mm
Supply Voltage	1.5 V
Power Consumption	6 mW

$1/f = 12dB$, which corresponds to frequency 64.5 MHz. It can be seen from the figure that the phase margin is over 80°.

The simulated transient response of the opamp is shown in Figure 6.13. It can be seen that the opamp output settles to within 0.2% error in less than 15ns.

The dynamic performance of the ADC can be characterized by its signal-to-noise-plus-distortion ratio (SNDR) and spurious-free dynamic range (SFDR). As described in Chapter 3, SNDR is defined as the ratio of the signal power to the total noise plus distortion power at the ADC output within a specified frequency band. It can be calculated from the power spectrum of the ADC output. Figure 6.14 shows the simulated power spectrum of the ADC output with an 1.1MHz input signal sampled at 20MS/s. The SNDR is found to be 45dB. Considering the limited accuracy of the simulator, the ADC performance basically reaches our design goal.

This ADC operates at a supply voltage of 1.5V, and consumes only 6mW when sampled at 20MHz.

The performance of the ADC is summarized in Table 6.3.

9. Summary

In this chapter, an 8-bit switched-opamp pipeline ADC is designed and implemented in a $0.18\mu m$ CMOS process for Bluetooth/WLAN(FHSS)/ HomeRF applications. This ADC operates at a single 1.5V power supply, and consumes only 6mW power when sampled at 20MS/s. It shows that the modified switched-opamp technique proposed in Chapter 5 is suitable for designing video-rate medium resolution analog circuits at very low voltages, without the need of on-chip clock boosting circuits or low-V_t devices.

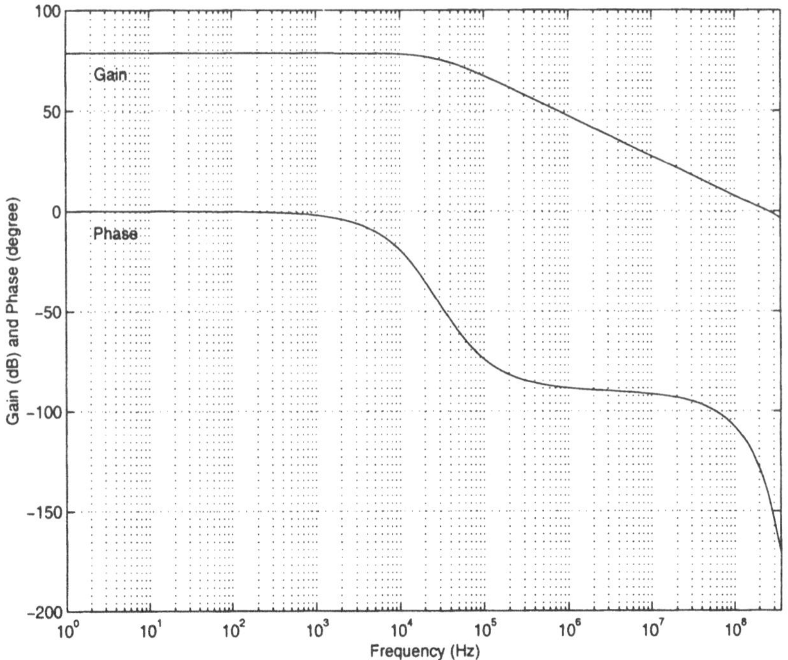

Figure 6.12. Simulated Frequency Response of Opamp

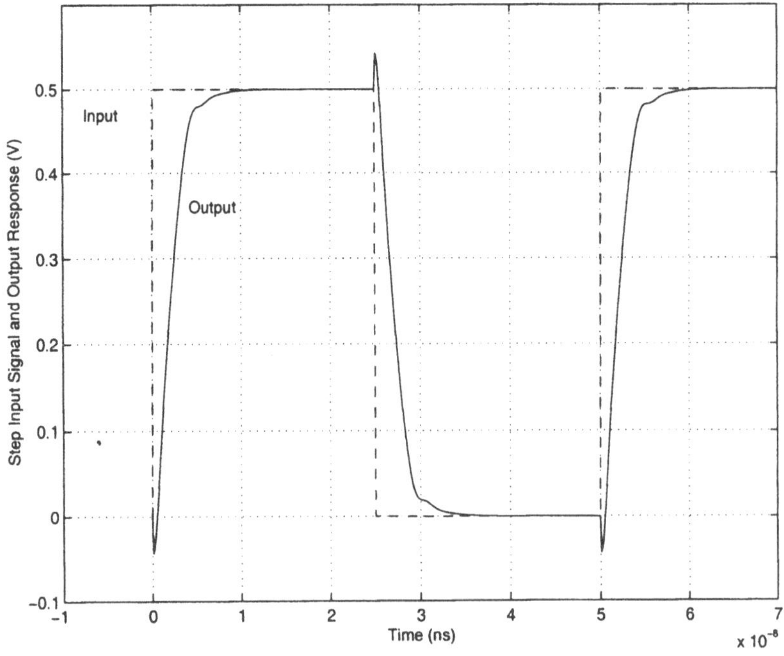

Figure 6.13. Simulated Transient Response of Opamp

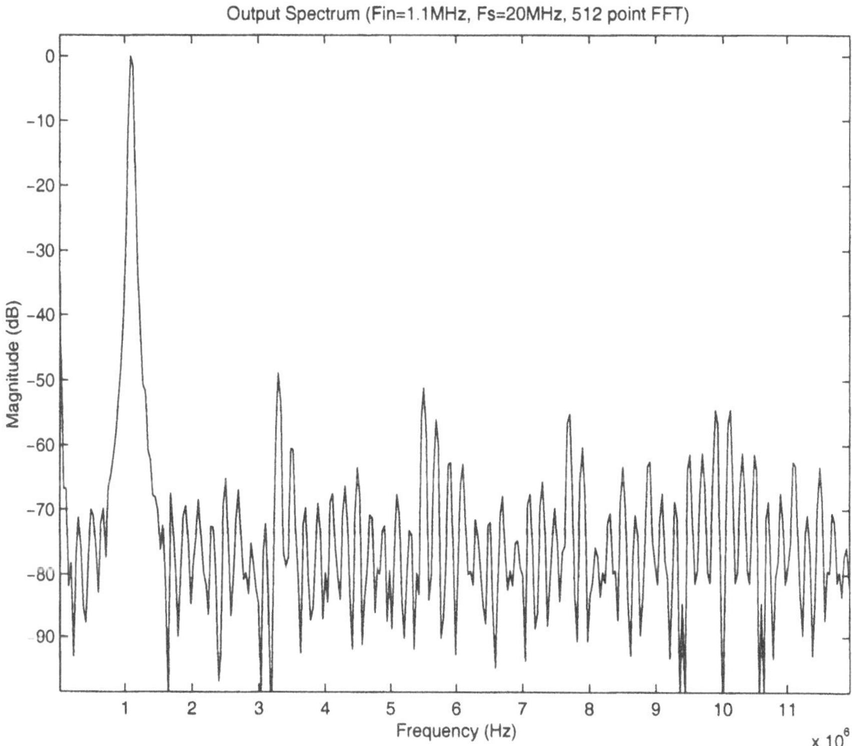

Figure 6.14. Power Spectrum of ADC Output

Chapter 7

HIGH-RESOLUTION DAC DESIGN TECHNIQUES

Digital-to-analog converter (DAC) is another essential building block in mixed-signal systems. This chapter focuses on the design techniques of low speed high resolution control DACs used in wireless transceivers. These DACs have different performance requirements from those used in the signal path. As discussed in Chapter 3, dynamic performances such as SFDR and SNDR are usually preferred when characterizing data converters used in the signal path. On the contrary, the conventional static performances such as INL and DNL are more useful when characterizing low speed DACs for control purposes.

This chapter first gives a brief review of three typical DAC architectures. Several design issues are then discussed, and novel techniques are proposed to improve the intrinsic accuracy of resistor-string DACs.

1. Review of DAC Architectures

This section is an overview of three commonly used DAC architectures: current steering DAC, switched-capacitor DAC, and resistor string DAC. The resistor string DAC is determined to be the most suitable architecture for low-power low-cost control applications.

1.1 Current Steering DAC

Current steering DACs [Miki et al., 1986, Schouwenaars et al., 1988, Cremonesi et al., 1989, Pelgrom, 1990, Bastiaansen et al., 1991, Nakamura et al., 1991, Fournier and Senn, 1991, Lin and Bult, 1998, Bastos et al., 1998, Van den Bosch et al., 2000, Van der Plas et al., 1999] have been extensively used in video, digital TV, HDTV and broad-band communication applications. The resolution can be up to 14-bit without calibration [Van der Plas et al., 1999], or 16-bit if calibration is employed [Schouwenaars et al., 1988]. The sampling rate can be up to 500MS/s [Lin and Bult, 1998].

Current steering DACs are based on an array of matched current cells that are steered to the DAC output depending on the digital input codes [Van der Plas et al., 1999]. The current cells are usually organized as a segmented architecture which combines an unary encoded current cell matrix as shown in Figure 7.1, and some binary weighted current cell elements. In this architecture, the least significant bits (LSBs) steer binary weighted current cells, while the most significant bits (MSBs) are thermometer encoded and steer the current cell matrix.

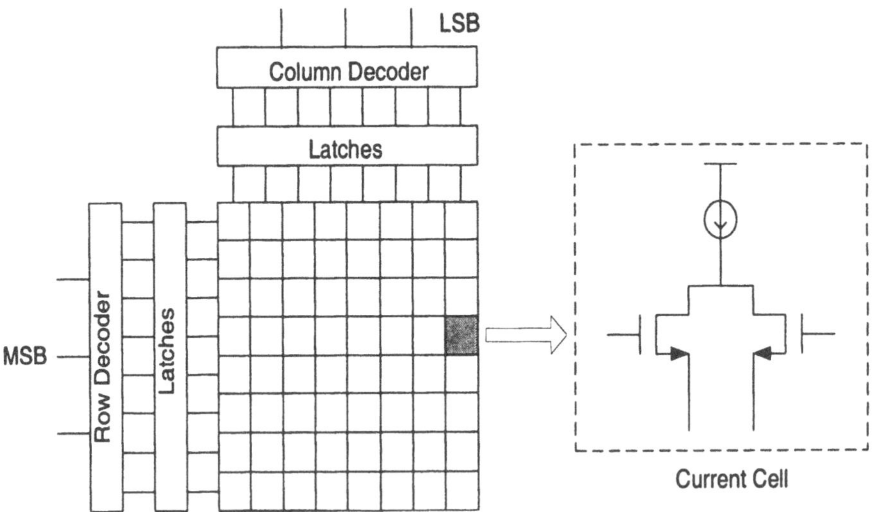

Figure 7.1. A Typical Current-Steering DAC Based on Unary Current Cell Matrix

Current steering DACs are intrinsically faster and more linear than other architectures. They can also directly drive an output resistive load without an extra output buffer. Furthermore, they can be easily integrated in fully digital CMOS technologies, making them a suitable choice for high-speed mixed-signal applications.

However, when operating at low sampling rates, current steering DACs are not competitive compared to other architectures. They tend to consumes more power and occupy more area than resistor string DACs.

1.2 Switched-Capacitor DAC

If we replace the input capacitor of a switched-capacitor gain amplifier by a programmable capacitor array (PCA) of binary-weighted capacitors, we can get a switched-capacitor DAC [Jones and Martin, 1996]. Consider a typical switched-capacitor DAC shown in Figure 7.2 [Lynn and Ferguson, 1997]. During phase ϕ_2, sampling capacitors are charged to V_{ref} or V_{ss} based on digital input codes. During phase ϕ_1, charges on sampling capacitors are integrated onto the integrating capacitor, generating an output proportional to the input code.

In today's standard mixed-signal IC technologies, polysilicon capacitors provide the best component matching per unit of area [Lynn and Ferguson, 1997]. Consequently, the use of switching capacitors in DACs provides excellent matching without sacrificing die areas. It also allows for very low-power operations.

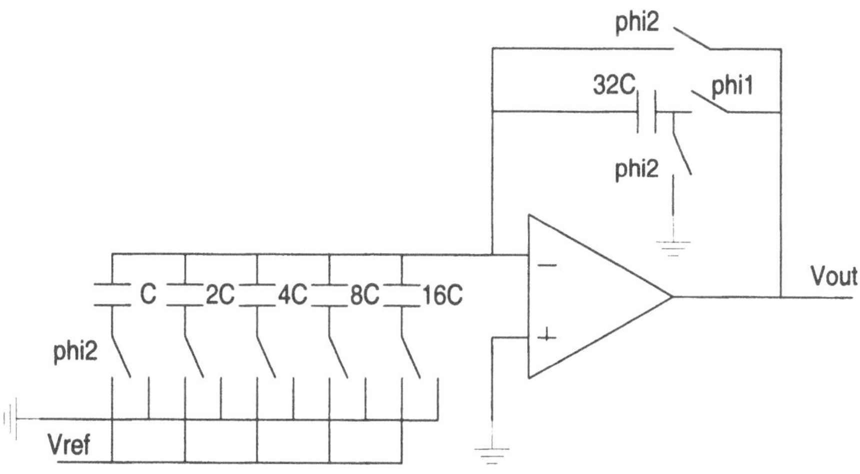

Figure 7.2. A Basic Switched-Capacitor DAC

The major drawback with switched-capacitor DACs is their inherent switching noise and large glitch energies. As a consequence, it is not advisable to use them to control RF blocks in wireless transceivers, since a small glitch at the output of a control DAC may easily destroy the performance of a whole transceiver.

1.3 Resistor String DAC

Another widely used DAC architecture [Jones and Martin, 1996, Oborn, 1996, Hamade, 1978] is resistor string converters. In this architecture, a voltage reference is applied to a resistor string, and the tap voltage is selected by a switch network, followed by a voltage buffer, as shown in Figure 7.3 [Jones and Martin, 1996]. Resistor string DACs have been extensively used in integrated circuits for several decades. Some variations including folded resistor string architecture [Abrial et al., 1988] and multi-resistor-string architecture [Holloway, 1984] have also been proposed to reduce the amount of digital coding circuits.

The popularity of resistor string DACs is based on its simplicity, inherent monotonicity and the continuous-time voltage output with low glitch energies. Compared to the current steering architecture, resistor string DACs have power and area savings when working at low speeds. They are also more suitable for control applications than the switched-capacitor architecture, since the latter has larger switching noise and glitch energy.

2. Intrinsic Matching of Resistor-String DAC

The static performance of resistor string DACs is strongly dependent on the intrinsic matching property of unit resistors, which is in turn determined by the process used. In a modern digital-oriented CMOS process, the matching between two resistors is typically at $8 \sim 10$-bit level. In order to achieve higher than 10-bit linearity, certain type of trimming (e.g. laser trimming) or calibration is usually required, which will be either too costly or need additional area and power.

This section first reviews the causes of nonlinearities in a single resistor string. Novel techniques for implementing resistor string DACs which compensates random and gradient errors are then presented as proposed by author in [Shi et al., 2001a].

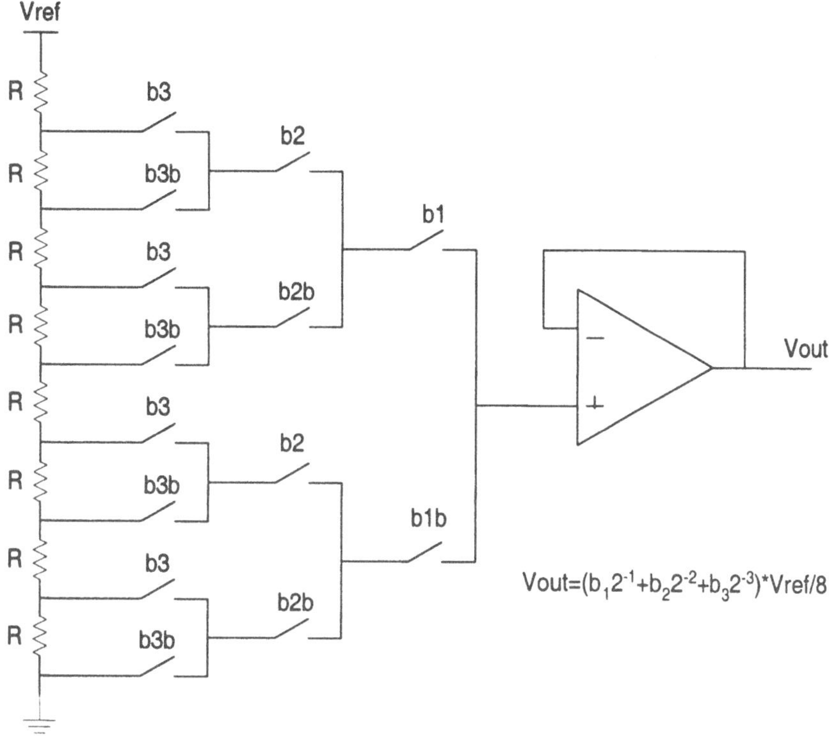

Figure 7.3. A 3-bit Resistor-String DAC

2.1 Resistor Matching Model

In a typical resistor string as shown in Figure 7.4, the nonlinearity comes from various sources.

1). Random error: caused by resistor mismatches. It can be shown [Kuboki and *et al.*, 1982] that tap voltages $V_1, V_2, ... V_N$ follow a nearly Gaussian distribution, and the maximal INL is given by:

$$INL_{max} = \frac{1}{\sqrt{4N}} \frac{\Delta R}{R} V_{ref} \tag{7.1}$$

2). Gradient errors: including thermal, doping and oxide thickness gradients. In general, wafers exhibit a linear pattern in the oxide thickness [Pelgrom et al., 1989], while temperature and stress gradients have a parabolic pattern [Bastos et al., 1997]. If only linear gradient error is

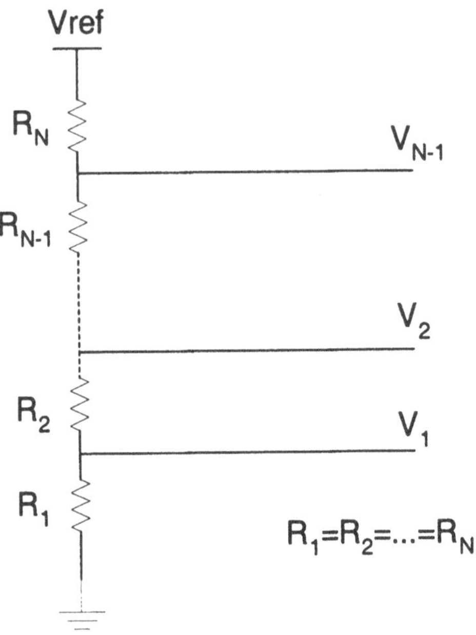

Figure 7.4. A Single Resistor String

considered, then the INL of the above resistor string can be derived as [Razavi, 1995]:

$$INL_{max} = N\frac{\Delta R}{8R}V_{ref} \tag{7.2}$$

As can be seen from (7.1) and (7.2), to minimize the overall INL, the resistor mismatch $\frac{\Delta R}{R}$ should be minimized. Without the aid of trimming or calibration, the previous art of improving matching is to simply increase the area of each unit resistor. It is not area efficient since the active area of the resistor string has to increase by a factor of four for each extra bit of DAC accuracy [Cong and Geiger, 2000]. And more importantly, this method only takes into account random mismatch. In the reality, when the dimension of the resistor string increases to a certain limit, the linear and parabolic gradient errors begin to dominate, so any further increase in size will not improve the linearity.

2.2 Design Techniques for Improved Resistor Matching

The static performance of DACs can only be improved by tackling all possible random and gradient errors [Van der Plas et al., 1999]. In this section, several design techniques suitable for resistor string DACs are proposed.

2.2.1 Reducing Random Errors

Figure 7.5 shows a new connection of a unit resistor. Instead of simply increasing the dimensions, the unit resistor is divided into M (=4 in this example) equal size, parallel-connected sub-resistors. For the connection with $M = 4^N$, N extra bits of accuracy can be obtained.

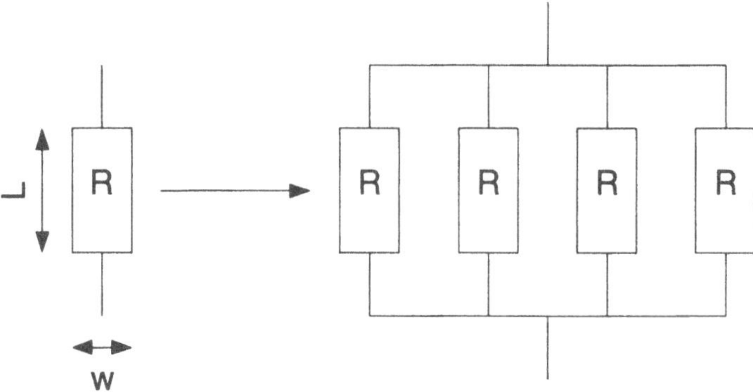

Figure 7.5. Dividing an Unit Resistor into 4 Parallel Sub-Resistors

The proof of this approach can be derived as follows. Assuming the resistance of each unit resistor R can be written as $R = R_0(1 + \epsilon)$, where R_0 is the nominal resistance and ϵ is the relative error, which follows Gaussian distribution: $\epsilon \sim G(0, \sigma)$, then the total resistance of four parallel-connected unit resistors can be derived as

$$
\begin{aligned}
R &= R_0(1+\epsilon_1)//R_0(1+\epsilon_2)//R_0(1+\epsilon_3)//R_0(1+\epsilon_4) \\
&= R_0 \frac{1}{\frac{1}{1+\epsilon_1} + \frac{1}{1+\epsilon_2} + \frac{1}{1+\epsilon_3} + \frac{1}{1+\epsilon_4}} \\
&\approx R_0 \frac{1}{1-\epsilon_1+1-\epsilon_2+1-\epsilon_3+1-\epsilon_4}
\end{aligned}
$$

$$\approx \frac{R_0}{4}(1 + \frac{\epsilon_1 + \epsilon_2 + \epsilon_3 + \epsilon_4}{4}) \tag{7.3}$$

It is well known that if $\epsilon \sim G(0, \sigma)$, then $\frac{\epsilon_1+\epsilon_2+\epsilon_3+\epsilon_4}{4} \sim G(0, \sqrt{4}\sigma/4) = G(0, \sigma/2)$. This means that parallel connecting four sub resistors can increase the DAC accuracy by one bit.

One disadvantage of this architecture is that the effective resistance is now R_0/M, which in turn increases power consumption of the resistor string by a factor of M. To solve this problem, a modified architecture as shown in Figure 7.6 can be used. This serial-parallel connection keeps the effective resistance unchanged, while the accuracy is still improved to the same extent as in Figure 7.5. Actually this structure is similar to doubling both W and L of the resistor, while gives more freedom in layout. This characteristic is useful for reducing gradient errors as will be explained in the next section.

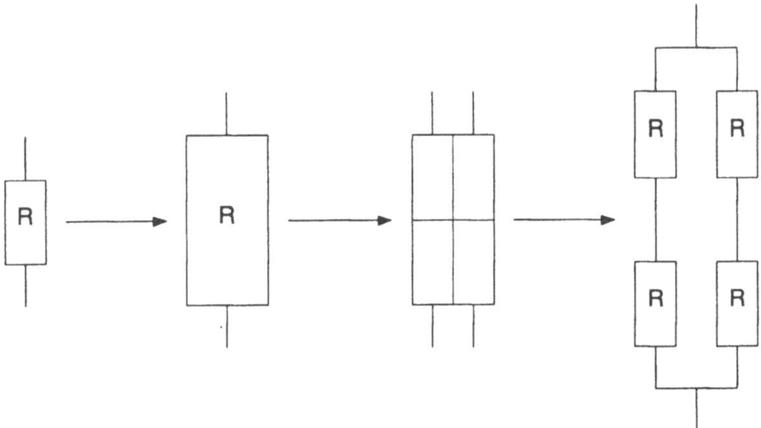

Figure 7.6. Modified Unit Resistor Connection

To illustrate the effectiveness of above structures, a Monte Carlo simulation is used to estimate the design yield under different unit resistor standard deviations. Figure 7.7 gives the simulation results on an 11-bit accuracy resistor string DAC. From the graph, one concludes that for a 99% yield, σ_R/R smaller than 0.1% is required for a normal resistor string, while the requirement is relaxed to 0.2% and 0.4% for 4 and 16 parallel-connected resistor strings, respectively.

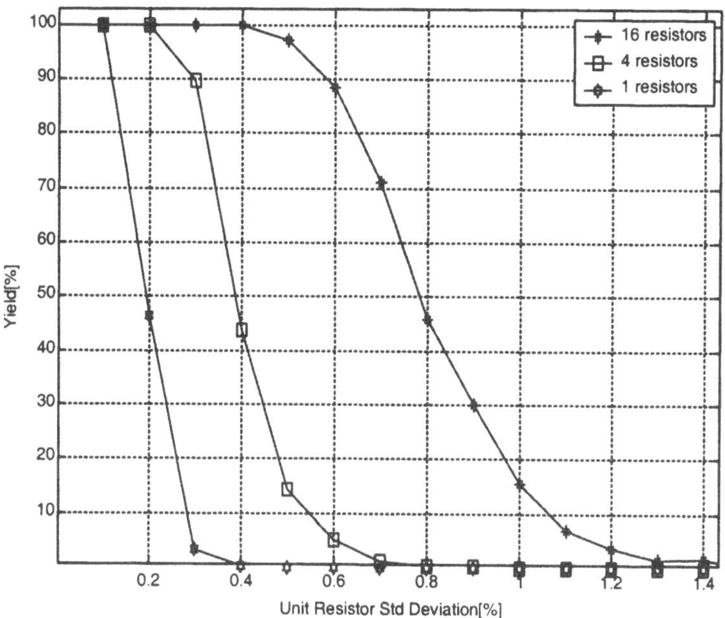

Figure 7.7. Yield vs. Unit Resistor Standard Deviation

2.2.2 Reducing Gradient Errors

Besides the random error, one should also consider gradient errors, especially when the dimension of the resistor string is large, as in the case of high accuracy DACs. The gradient error distribution in a resistor array can be approximated by the linear and parabolic terms in a Taylor series expansion and expressed as

$$\epsilon(x, y) \approx a_0 + a_{11}x + a_{12}y + a_{21}x^2 + a_{22}xy + a_{23}y^2 \qquad (7.4)$$

where (x,y) is the coordinate of the unit in the resistor array.

Special layout techniques can be used to reduce these errors. Figure 7.8 shows an example of a 4-bit resistor string layout and explains how the techniques discussed in the previous section can be employed. Assume the 4-bit resistor string is laid out in a 4×4 array, as shown in Figure 7.8(a). If we divide each unit resistor into 4 sub resistors, the resistor array can be mirrored twice and the layout as shown in Figure 7.8(b) can be used.

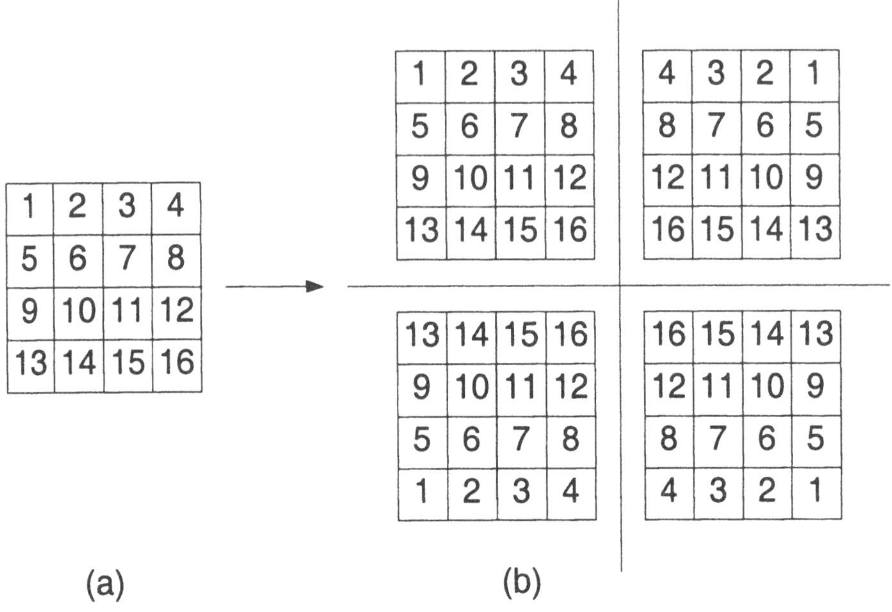

Figure 7.8. Linear Error Compensated Layout

This layout scheme suppresses the linear gradient errors since it is symmetrical in both $x-$ and $y-$ directions. The remaining parabolic gradient errors can be further reduced by "random walk" [Van der Plas et al., 1999] or "best INL" [Cong and Geiger, 2000] layout scheme introduced in current steering DACs. Figure 7.9 illustrates the "best INL" layout of the same 4-bit resistor string. The basic idea of this technique is to randomize the layout in each resistor array, which helps prevent the accumulation of residue errors.

As an example, Figure 7.10 gives the INL (refer to 11bit accuracy) of a 4-bit resistor string under different layout arrangements. As can be seen, under the presence of both linear and parabolic gradient errors, the 2×2 symmetrical layout reduces the maximal INL by approximately one half, and the "best INL" scheme further reduces the INL.

In summary, by dividing the unit resistor into parallel or serial-parallel sub resistors and employing special layout schemes, both the random and

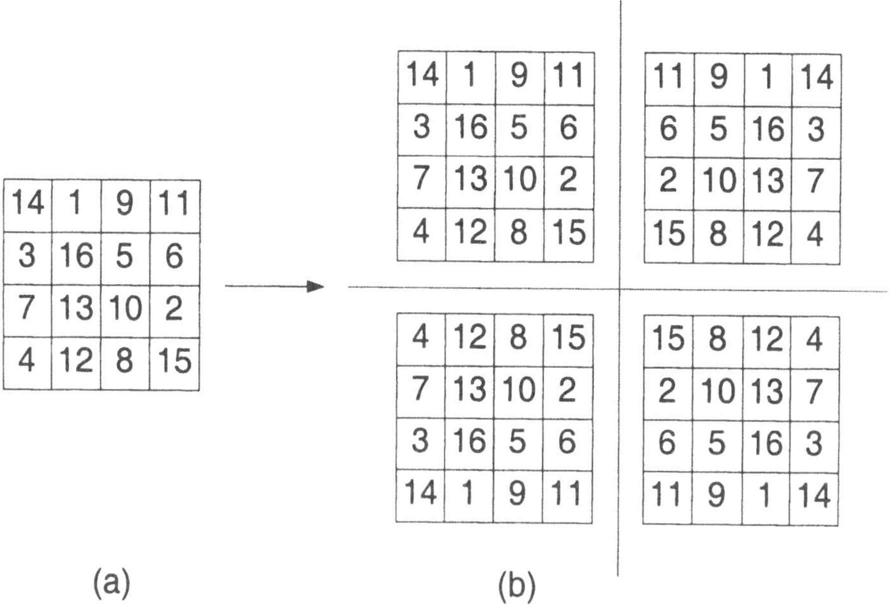

Figure 7.9. Modified Layout Arrangement

gradient errors can be minimized, thus the accuracy of the resistor string can be substantially improved.

Note that in the actual layout, resistors are usually laid out in a polysilicon layer, but all the contacts and metal connections have additional resistance. Care must be taken to match the number of contacts and the length of wires in order to achieve the best accuracy.

3. Summary

In this chapter, novel techniques for improving intrinsic accuracy of resistor string DACs without trimming, calibration or dynamic averaging are proposed. By dividing the unit resistor into multiple parallel-connected sub-resistors and arranging them in a symmetrical but randomized manner, the DAC accuracy can be improved by up to 2-bits with less than 25% area increase.

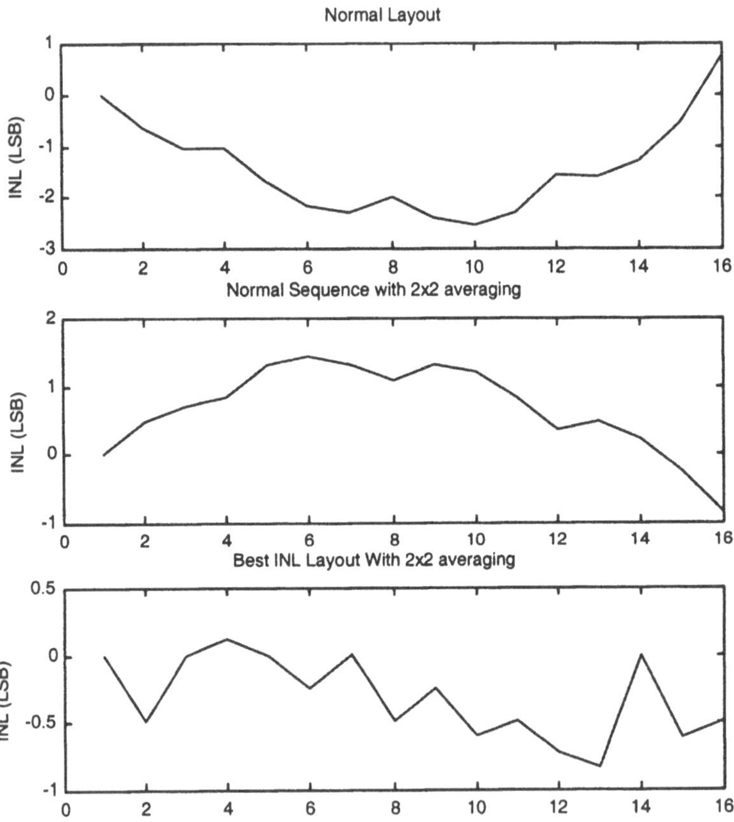

Figure 7.10. INL of Different Layout Schemes

Chapter 8

CONTROL DAC FOR 3G (UMTS) TRANSCEIVERS

In wireless transceivers, a large number of control DACs are used as auxiliary circuits to control RF blocks or baseband VGAs. For cost reasons, it is preferable for them to be integrated with digital circuits in deep submicron CMOS processes. However, in a typical digital process, the resistor matching is limited to 8~10-bit accuracy. The design techniques proposed in Chapter 7 can then be employed to implement the DAC with up to 11-bit accuracy without trimming or calibration.

In this chapter, the design of an 11-bit resistor string DAC is described. The design details at both the system and the block level are given in the following sections.

1. Applications

The second-generation (2G) cellular systems have shown great success worldwide due to the effective utilization of digital technologies to increase capacity, improve reliability and lower system size and cost. However, these systems are mainly limited to voice and low data rate services.

With the standardization of the third generation (3G) cellular systems, wireless communications will evolve from voice service to high data rate multimedia applications. The 3G system, also known as UMTS (Universal Mobile Telecommunication System) will provide universal access and global roaming. The IMT-2000 (International Mobile Telecommunication 2000) standard is intended to unify the diverse 2G systems and form a basis for the third generation systems. The 3G systems provide a wide range of services from voice, internet access, to high speed videoconference. The data rate can be up to 2Mbps in picocell environment.

The prototype DAC is designed for control purpose in 3G transceivers. It is employed to control RF or baseband components. For example, in order to fully utilize the input range of the ADC, a control DAC can be used to adjust the gain of the baseband VGA based on the input signal strength. The control signal (digital codes) is usually generated by a DSP or microprocessor.

2. Design Specifications

These control DACs typically have low sampling rates ($< 100kS/s$) and $6 \sim 11$ bit linearity, but require low power and small area for cost savings. Table 8.1 lists design specifications of the prototype DAC.

Table 8.1. Specifications of the Prototype DAC

Technology	$0.18\mu m$ CMOS
Resolution	$10 \sim 11$-bits
Conversion Rate	$\leq 10KS/s$
Load Resistance	$\geq 10K\Omega$
Load Capacitance	$\leq 50pF$
Supply Voltage	3.0 V

Note that both resistive and capacitive loads are presented at the DAC output, making this DAC suitable for various control purposes.

3. Architecture

For DACs with 10-bit or higher resolutions, it is not practical to implement them with a single resistor string due to its prohibitively large area and load capacitance.

A folded resistor string can be used to reduce the capacitive loading, as shown in Figure 8.1 [Abrial et al., 1988]. This approach reduces the total number of transistor junctions on the output line to $2\sqrt{2^N}$ from 2^N as in the case of conventional single resistor string DACs. Unfortunately, the number of resistors is still the same as the previous approach, implying a large chip area.

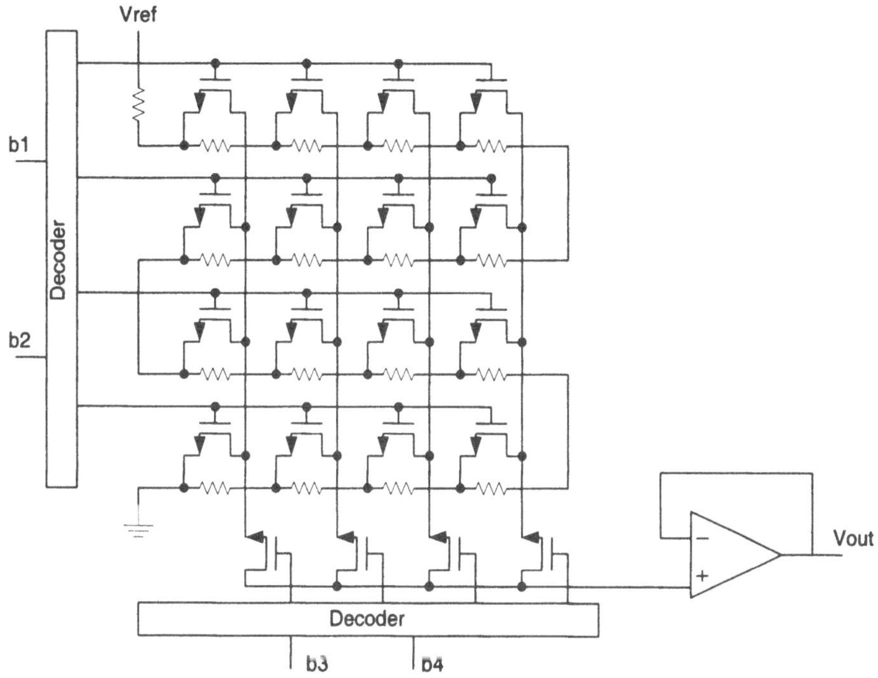

Figure 8.1. A 4-bit Folded Resistor String DAC

Multiple resistor string architecture was first proposed in [Holloway, 1984] to reduce the number of resistors. In this architecture as shown in Figure 8.2, the MSBs codes determine which two adjacent nodes of the first resistor string are connected to the two voltage buffers. The final output is determined by the lower LSBs codes and the second resistor string connected between these two buffers.

This architecture is suitable for high resolution applications. However, since it introduces two additional buffers, the monotonicity of the DAC depends on the matching of the input offset voltages between these two buffers. In CMOS processes, the offset voltage of an opamp is usually in the range of $5 \sim 20mV$ and is difficult to control; while in BiCMOS processes, opamps can be made with low noise and small input offsets. As a consequence, this architecture finds its most applications in

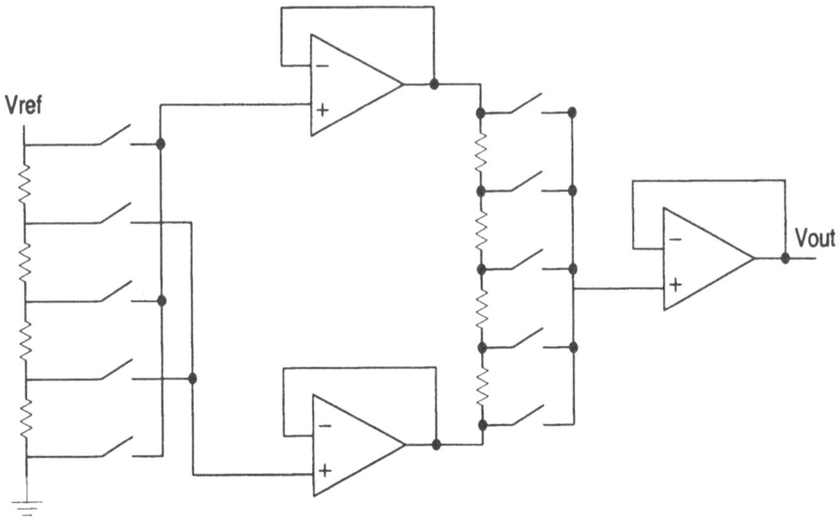

Figure 8.2. Multiple Resistor String DAC (4bit)

BiCMOS technologies. Besides this limitation, it also tends to consume large power, since those two intermediate buffers need to drive resistive loads.

A modified multiple resistor string architecture which overcomes above drawbacks is used in this design. As shown in Figure 8.3, the second resistor string is connected directly to the output nodes of the first string, without two intermediate buffers. Monotonicity is guaranteed in this architecture. In order to reduce the load effect of the second string on the first string, R_2 should be much larger than R_1. Since the second resistor string has much relaxed matching requirements, small size devices can be used in the layout. Consequently, this modified architecture only introduces limited area increase.

One problem with this architecture is that it reduces the maximal operating speed due to the large unit resistance R_2 of the second resistor string. To solve this, two small current sources can be added to the second resistor string, as shown in Figure 8.4. The current I_0 is set to be $\frac{V_{ref}}{2^{n1}2^{n2}R_2}$, where n_1 and n_2 are the resolutions of the first and second resistor string, respectively. The basic idea of this approach is to provide an external current to the second resistor string. It eliminates the load

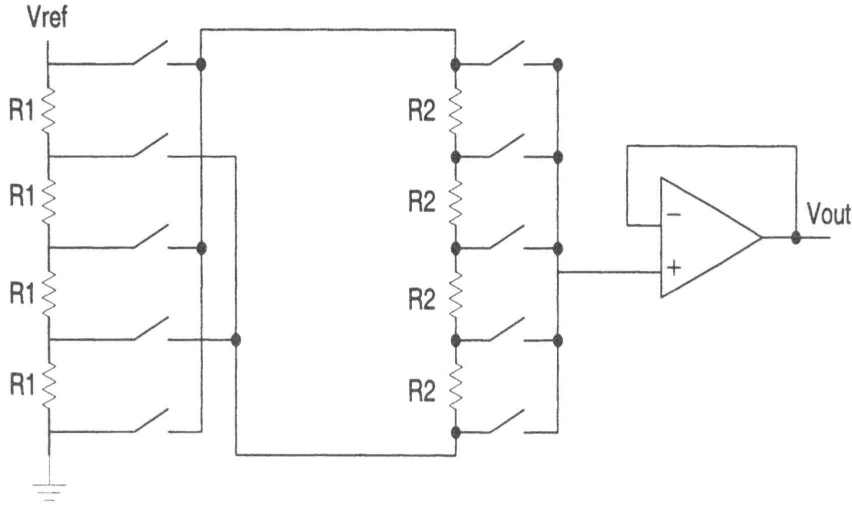

Figure 8.3. A Modified Multiple Resistor String DAC (4bit)

effect of the second string without using large resistors. As a result, it can operate at higher speeds.

However, since speed is not a major concern in control DACs, the architecture shown in Figure 8.3 is used in this design because of its simplicity.

4. Design of Resistor Strings

There are two resistor strings in this prototype DAC: a 4-bit MSB string and a 7-bit LSB string. The LSB string is connected directly to the MSB string, acting as a resistive load. To reduce this load effect, unit resistance R_2 of LSB string is chosen to be much large than R_1 of MSB string. As an example, Figure 8.5 shows the simulated nonlinearity performance of the DAC when $R_2 = 8R_1$. As can be seen, the integral nonlinearity (INL) is between -0.18LSB and +0.18LSB, and the differential nonlinearity (DNL) is very close to zero at most input codes, expect some spikes (about 0.25LSB) at major transition codes.

In order to meet the 11-bit accuracy requirement, the 4-bit MSB string should have 11-bit accuracy, while the 7-bit LSB string only needs 7-bit accuracy. If we assume that in a given process the best resistor matching

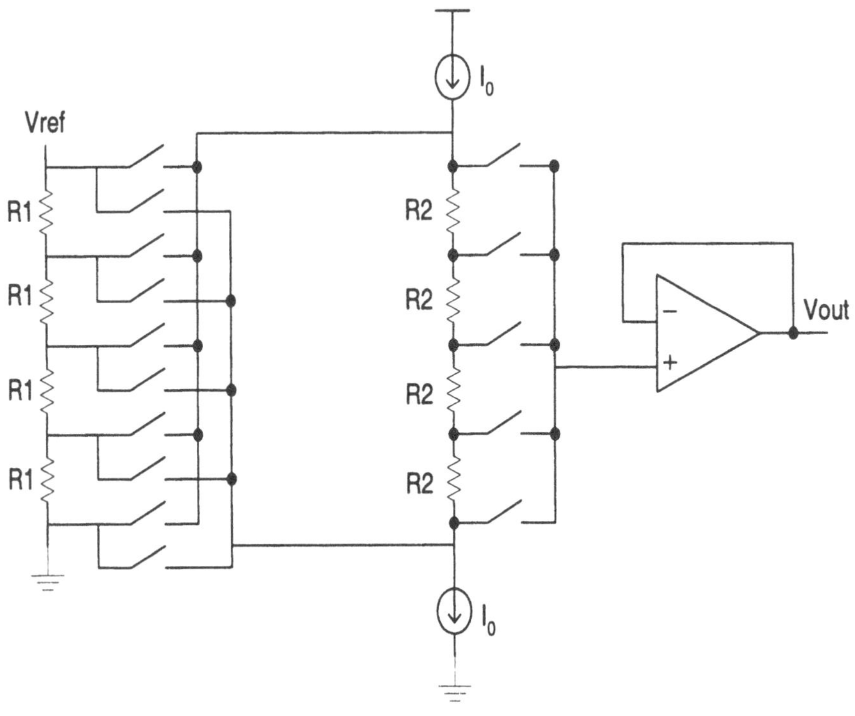

Figure 8.4. A Multiple Resistor String DAC with Two Compensation Currents

is in 10-bit level, then we need only to apply the random and gradient errors compensation schemes discussed in the previous section to the MSB string. The example shown in Section 2 (with $M = 4$) gives one extra bit of accuracy. If we further divide the unit resistor into 16 sub resistors and employ similar layout schemes, two extra bits of accuracy can be obtained (however, it is not encouraged to use $M > 16$ due the layout and routing difficulties). Since the MSB string has only $2^4 = 16$ resistors and the unit resistance is much smaller than that of LSB string, the total area penalty caused by the new layout scheme is small (usually $< 25\%$). And it is worthy to note that the proposed architecture consumes no additional power compared to a conventional resistor string DAC.

5. Class-AB Output Buffer

In the resistor string DAC, an extra output buffer is needed in order to drive heavy loads. The buffer used in this design is a two stage amplifier

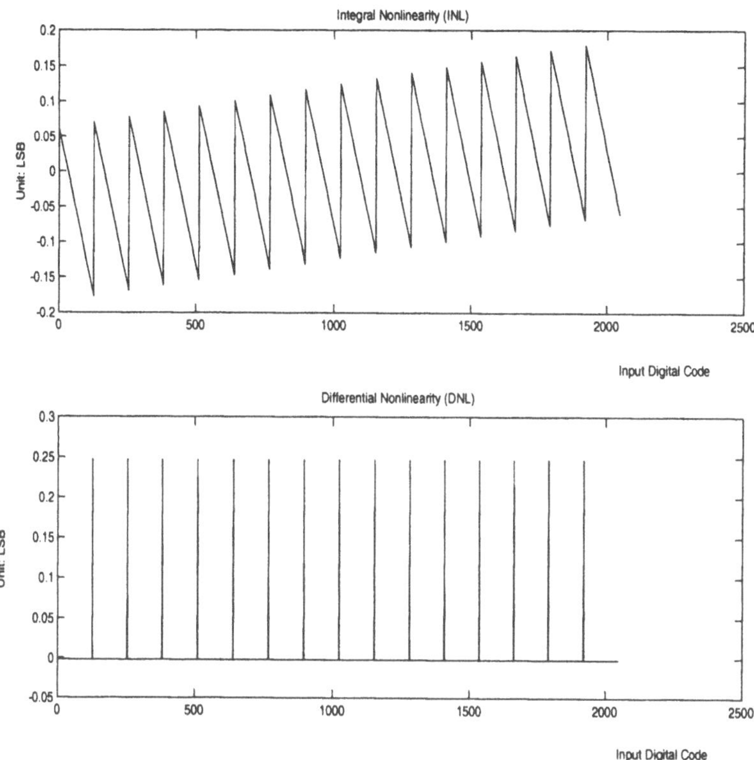

Figure 8.5. The DAC Nonlinearity Caused by the Load Effect of the LSB String

with a class-AB output stage (see Figure 8.6). The output stage consists of push-pull devices M_{19}, M_{20} with floating class-AB control devices M_9, M_{10} [Gregorian, 1999, Hogervorst et al., 1994]. At quiescent conditions, the current $2I$ flowing through devices M_7, M_8, M_{11}, and M_{12} is equally divided between devices M_9 and M_{10}. The quiescent current I_o in the output transistors is determined by two translinear loops M_{10}, M_{20}, M_{17}, M_{18} and M_9, M_{19}, M_{14}, M_{13}. Assuming $(W/L)_{10} = (W/L)_{17}$ and $(W/L)_9 = (W/L)_{14}$, the I_o can be expressed as

$$I_o = I \times \frac{(W/L)_{20}}{(W/L)_{18}} = I \times \frac{(W/L)_{19}}{(W/L)_{13}} \tag{8.1}$$

The class-AB action is performed by keeping the voltage between the gates of the output transistors constant. When the intermediate voltage

V_1 increases, the current I_{M10} decreases while I_{M9} increases by the same amount. As a result, both $V_{G,M20}$ and $V_{G,M19}$ move up. Thus the output stage pulls more current from the output node. A similar discussion can be held when V_1 decreases.

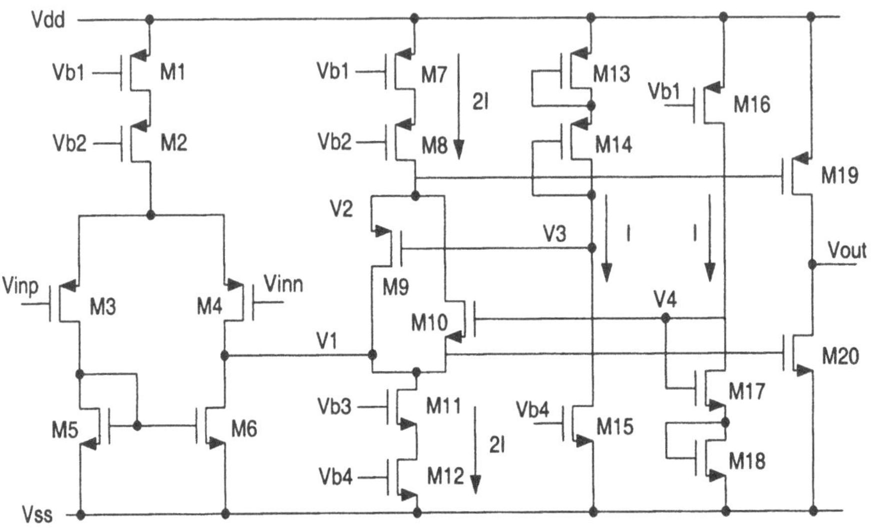

Figure 8.6. The Output Voltage Buffer Used in the DAC

The output buffer is designed to drive a 10KΩ resistive load and a 50pF capacitive load. Figure 8.7 shows the simulated AC response. The open loop DC gain is over 90dB. The unit gain frequency is 4.2MHz, and the phase margin is 72°. The buffer operates at a single 3V power supply, and consumes only $150\mu A$ quiescent current.

6. Deglitching Circuit

Due to the timing mismatch of digital control signals, there are always some glitches or spikes at the DAC output. Glitches are a major problem for D/A converters, especially for control DACs used in wireless transceivers. In multiple resistor string DACs, worst case glitches happen during the major code transitions. For example, in the prototype 11-bit DAC with a 4/7 architecture, large glitches occur at multiples of the 127 to 128 code transitions. At other codes, the glitch is negligible because of the first order R-C filtering effect (R: resistors in the LSB string; C: input capacitance of the output buffer).

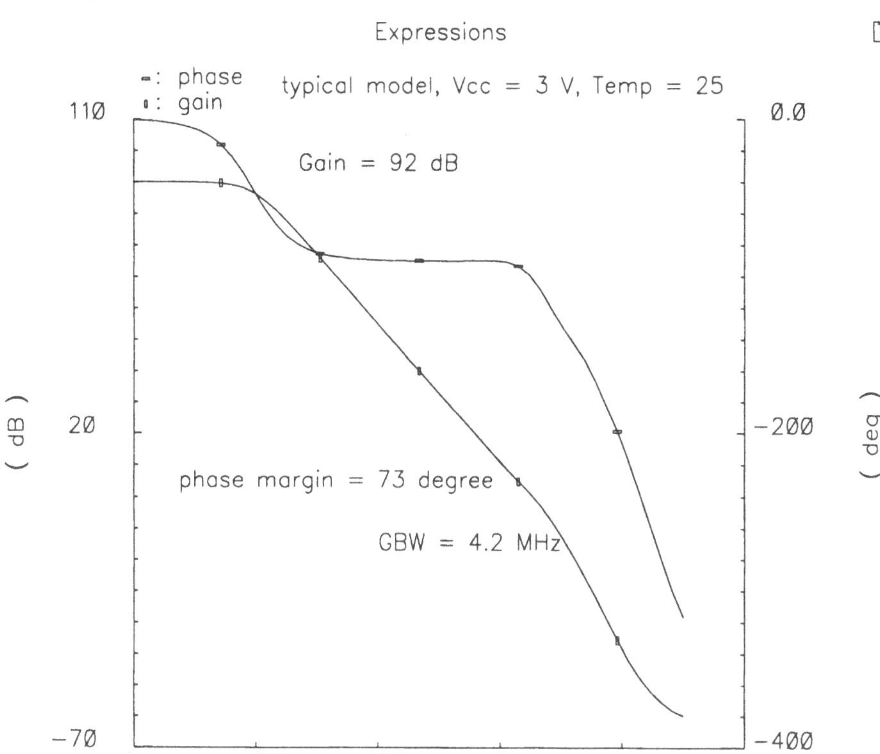

Figure 8.7. Open Loop Gain and Phase of the Output Voltage Buffer

To reduce the glitch energy, it is desirable to synchronize the control signals of the switching transistors for all input codes. However, it is difficult to keep the synchronization between MSB logics and LSB logics. In this section, a novel deglitching circuit is proposed which greatly reduces the glitches during major code transitions.

Figure 8.8 illustrates the block diagram of the proposed deglitching circuit. The underlying idea is to turn off the switch between the resistor string and the output buffer for a very short time when the major code transition occurs. The deglitching control block shown in Figure 8.8 is used to detect major transitions and generate the adequate control clock.

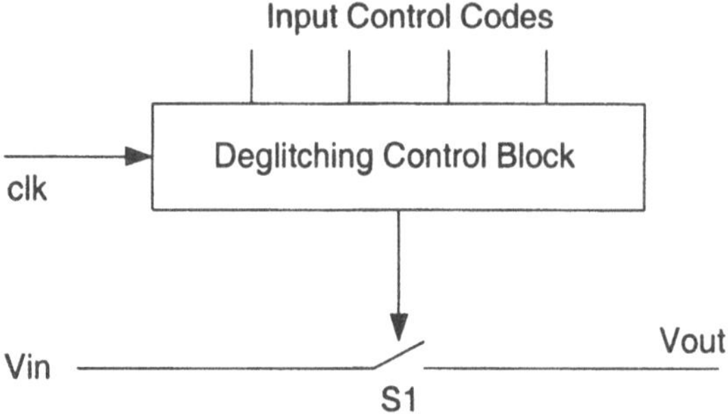

Figure 8.8. The Proposed Deglitching Circuit

This circuit is very effective in reducing major glitches at DAC outputs, and it does not affect the functionality of the DAC since the control switch S_1 is always on during normal code transitions. Furthermore, this circuit is very power and area efficient.

7. Performance of the Prototype DAC

The prototype 11-bit multiple resistor string DAC is designed and implemented with a $0.18\mu m$ CMOS process. The resistors are laid out in the polysilicon layer. The active die area is about $1\mu m \times 1\mu m$, among which the two resistor strings occupy about $0.5\mu m \times 0.5\mu m$.

As discussed in Chapter 7, static performances such as integral nonlinearity (INL) and differential nonlinearity (DNL) are often used to measure low speed control DACs. Monte-Carlo simulation is used to verify the performance of the DAC. Figure 8.9 shows the simulated INL and DNL performance. In the simulation, each unit resistor has a nominal resistance term, a random error term, a linear error term and a parabolic error term. All these error terms are further assumed to be small and independent, therefore the resistor mismatch error is the sum of all three error terms. As can be seen from Figure 8.9, the maximum INL is around 1LSB (least significant bit), and the the maximum DNL is less than 0.3LSB. The maximum DNL only happens during major code transitions (The input code changes from 127 to 128 and its multiples), while at other input digital codes, the DNL is close to zero.

Shown in Figure 8.10 is the transient response observed at the buffer output when the input digital code changes from 1024 to 1023 (one of

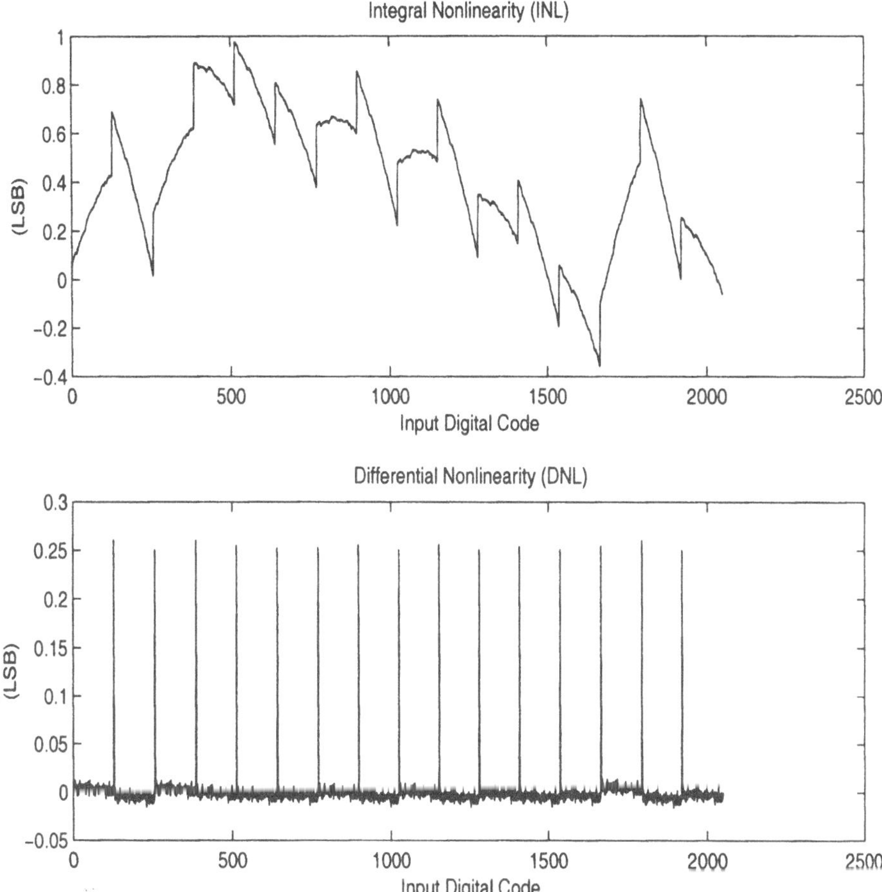

Figure 8.9. Simulated Nonlinearity of the DAC

major transitions). As can be seen, the glitch level is less than 0.5mV, corresponding to 1/2LSB. As a comparison, the major glitch is found to be more than 5mV if the deglitching circuit is not used.

It is also worthy to note that in Figure 8.10, the output voltage change is around 1.42mV (1.136LSB) when the input code changes from 1024 to 1023, which corresponds to the DNL spike at code 1024 as shown in Figure 8.9. However, unlike the Monte Carlo simulations, ideal resistors are used in glitch simulations, which is the reason why the DNL values from these two figures are different.

The DAC operates at a typical supply voltage of 3V, and the total static power consumption is only $750\mu W$.

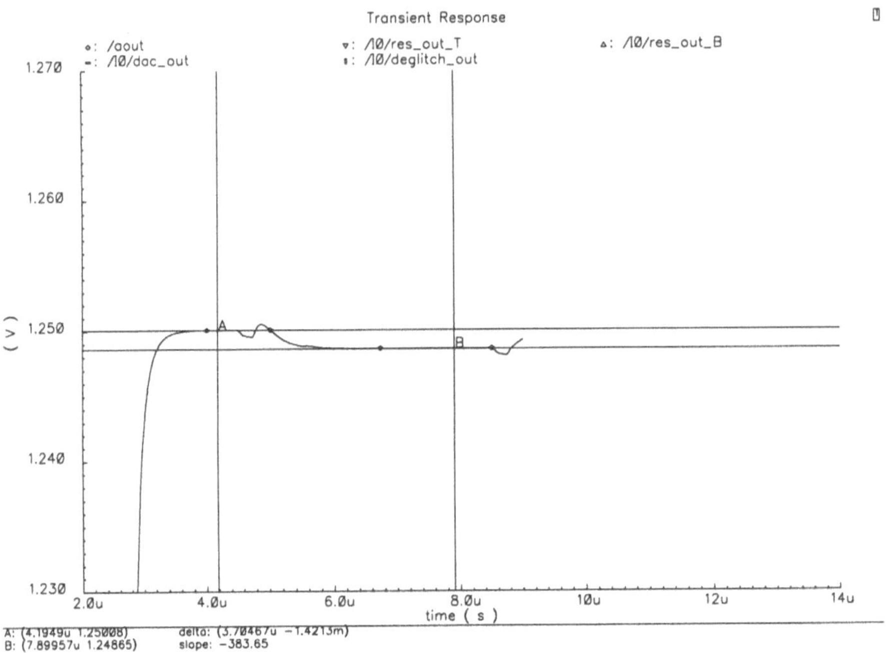

Figure 8.10. The Major Glitch at the DAC Output

The performance of the prototype DAC is summarized in Table 8.2.

8. Summary

An 11-bit resistor string DAC is designed and implemented in a $0.18\mu m$ CMOS process. The design techniques proposed in Chapter 7 are employed to design this high resolution DAC in a deep submicron digital process without trimming or calibration. A special deglitching circuit is proposed to reduce the DAC output glitches during major code transitions. Simulation results show that the prototype DAC has true 11-bit accuracy, and the proposed deglitching circuit reduces the level of major glitches by an order of magnitude.

Table 8.2. Performance Summary of the Resistor String DAC

Parameter	Values
Technology	$0.18\mu m$ CMOS
Resolution	11bits
Conversion Rate	100 KS/s
INL (max)	1 LSB
DNL (max)	0.3 LSB
Major Glitch Level	$<$ 0.5 mV
Resistive Load	$\geq 10K\Omega$
Capacitive Load	$\leq 50pF$
Active Area	1.0mm x 1.0mm
Supply Voltage	3.0 V
Power Consumption	750 μW

Chapter 9

CONCLUSION

In this book, design and power optimization techniques of CMOS data converters are investigated. The summary is as follows:

1. In pipeline ADCs, amplifier/capacitor scaling and digital error correction are two effective design techniques for power reduction. A novel approach named dynamic biasing is proposed. In this approach, the amplifier is partially turned off during the idle state. The effectiveness of this technique is proved in a prototype ADC design. This technique can also be applied to other ADC architectures.

2. Low voltage operation is another important key factor in portable electronic devices. It also poses severe challenges on the baseband circuit designs. Switched-opamp technique is more compatible with deep submicron technologies and suitable for system-on-chip designs than its counterpart techniques. Several modifications to the conventional switched-opamp technique are proposed to improve the sampling frequency. A experimental prototype is design and implemented in a $0.18\mu m$ CMOS process. This 8-bit pipeline ADC operates at a 1.5V supply, and dissipates only 6mW when sampled at 20MS/s.

3. A large number of control DACs are needed in modern wireless transceivers. Resistor string DAC is suitable for these applications due to its low power, small area and low glitch energy. A novel technique is proposed to improve the intrinsic accuracy of resistor string DACs without trimming and calibration. In this approach, the unit resistor is

divided into several serial-parallel connected sub resistors, and the entire resistor string is laid out in a symmetrical but randomized manner. This technique is suitable for high resolution multiple-resistor-string DACs.

References

[roa, 1999] (1999). *"International Technology Roadmap for Semiconductors"*. Semiconductor Industry Association.

[Abidi et al., 2000] Abidi, A. A., Pottie, G. J., and Kaiser, W. J. (2000). "Power-Conscious Design of Wireless Circuits and Systems". *Proceedings of IEEE*, 88(10):1528–1545.

[Abo, 1999] Abo, A. M. (1999). *"Design for Raeliablity of Low-voltage, Switched-capacitor Circuits"*. Ph.D. Dissertation, University of California, Berkeley.

[Abo and Gray, 1999] Abo, A. M. and Gray, P. R. (1999). "A 1.5-V 10-b 14.3-MS/s CMOS Pipeline Analog-to-digital Converter". *IEEE Journal of Solid-State Circuits*, 34(5):599–606.

[Abrial et al., 1988] Abrial, A., Bourier, J., Fournier, J.-M., Senn, P., and Veillard, M. (1988). "A 27-MHz Digital-to-Analog Video Processor". *IEEE Journal of Solid-State Circuits*, 23(12):1358–1369.

[Adachi et al., 1990] Adachi, T., Ishinawa, A., Barlow, A., and Takasuka, K. (1990). "A 1.4 V Switched-capacitor Filter". *Proc. CICC*, pages 8.2.1–8.2.4.

[Baltus and Dekker, 2000] Baltus, P. G. M. and Dekker, R. (2000). "Optimizing RF Front Ends for Low Power". *Proceedings of IEEE*, 88(10):1546–1559.

[Baschirotto and Castello, 1997] Baschirotto, A. and Castello, R. (1997). "A 1-V 1.8-MHz CMOS Switched-opamp SC Filter with Rail-to-rail Output Swing". *IEEE Journal of Solid-State Circuits*, 32(12):1979–1986.

[Baschirotto et al., 1998] Baschirotto, A., Castello, R., and Montagna, G. P. (1998). "Active Series Switch for Switched-opamp Circuits". *IEE Electron. Lett.*, 34:1365–1366.

[Baschirotto et al., 1994] Baschirotto, A., Castello, R., and Montecchi, F. (1994). "Design Strategy for Low-voltage SC Circuits". *IEE Electron. Lett.*, 30:378–379.

[Bastiaansen et al., 1991] Bastiaansen, C. A. A., Geoeneveld, D. W. J., Schouwenaars, H. J., and Termeer, H. A. H. (1991). "A 10-b 40-MHz 0.8-μm CMOS D/A Current-Output D/A Converter". *IEEE Journal of Solid-State Circuits*, 26(7):917–921.

[Bastos et al., 1998] Bastos, J., Marques, A. M., Steyaert, M. S. J., and Sansen, W. (1998). "A 12-Bit Intrinsic Accuracy High-Speed CMOS DAC". *IEEE Journal of Solid-State Circuits*, 33(12):1959–1969.

[Bastos et al., 1997] Bastos, J., Steyaert, M., Pergoot, A., and Sansen, W. (1997). "Influence of Die Attachment on MOS Transistor Matching". *IEEE Tran. Semiconduct. Manufact.*, pages 209–217.

[Brandt and Lutsky, 1999] Brandt, B. and Lutsky, J. (1999). "A 75mW 10b 20Msample/s CMOS Subranging ADC with 59dB SNDR". *ISSCC Dig. Tech. Papers*, pages 322–323.

[Castello and Tomasini, 1991] Castello, R. and Tomasini, L. (1991). "A 1.5 V High-performance Switched-capacitor Filters in BiCMOS Technology". *IEEE Journal of Solid-State Circuits*, 26(7):930–936.

[Chaudhury et al., 1999] Chaudhury, P., Mohr, W., and Onoe, S. (1999). "The 3GPP Proposal for IMT-2000". *IEEE Communication Magazine*, 37(12):72–81.

[Cho and Gray, 1995] Cho, T. B. and Gray, P. R. (1995). "A 10b, 20Msample/s, 35mW Pipeline A/D Converter". *IEEE Journal of Solid-State Circuits*, 30(3):166–172.

[Cline, 1996] Cline, D. W. (1996). *"Noise, Speed, and Power Trade-offs in Pipelined Analog to Digital Converters"*. Ph.D. Dissertation, University of California at Berkeley.

[Cong and Geiger, 2000] Cong, Y. and Geiger, R. L. (2000). "Switching Sequence Optimization for Gradient Error Compensation in Thermometer-Decoded DAC Arrays". *IEEE Transactions on CAS-II*, pages 585–595.

[Conroy et al., 1993] Conroy, C., Cline, D., and Gray, P. R. (1993). "An 8-b 85-MS/s Parallel A/D Converter in 1-μm CMOS". *IEEE Journal of Solid-State Circuits*, 28(4):447–454.

[Cremonesi et al., 1989] Cremonesi, A., Maloberti, F., and Polito, G. (1989). "A 100-MHz CMOS DAC for Video-Graphic Systems". *IEEE Journal of Solid-State Circuits*, 24(6):635–639.

[Crols and Steyaert, 1994] Crols, J. and Steyaert, M. (1994). "Switched-Opamp: An Approach to Realize Full CMOS Switched-Capacitor at Very Low Power Supply Voltages". *IEEE Journal of Solid-State Circuits*, 29(8):936–942.

[Dessouky and Kaiser, 2001] Dessouky, M. and Kaiser, A. (2001). "Very Low-Voltage Digital-Audio $\Delta\sum$ Modulator with 88-dB Dynamic Range Using Local Switch Bootstrapping". *IEEE Journal of Solid-State Circuits*, 36(3):349–355.

[Dickson, 1976] Dickson, J. F. (1976). "On-chip High-voltage Generation in NMOS Integrated Circuits using an Improved Voltage Multiplier Technique". *IEEE Journal of Solid-State Circuits*, SC-11(6):374–378.

[Elwan et al., 2001] Elwan, H., Alzaher, H., and Ismail, M. (2001). "A New Generation of Global Wireless Compatibility". *IEEE Circuits and Devices Magazine*, 17(1):7–19.

[Fournier and Senn, 1991] Fournier, J. M. and Senn, P. (1991). "A 130-MHz 8-b CMOS Video DAC for HDTV Applications". *IEEE Journal of Solid-State Circuits*, 26(7):1073–1077.

[Gee, 2001] Gee, T. (2001). "Supressing Errors in W-CDMA Mobile Devices". *Communication System Design*, 7(3):33–38.

[Gendai et al., 1991] Gendai, Y., Komatsu, Y., Hirase, S., and Kawata, M. (1991). "An 8-b 500-MHz ADC". *IEEE Int. Solid-State Citcuits Conf.*, pages 172–173.

[Gray and Meyer, 1995] Gray, P. R. and Meyer, P. G. (1995). "Future Directions in Silicon ICs for RF Personal Communications ". *Proceedings of the IEEE 1995 Custom Integrated Circuits Conference*, pages 83–90.

[Gregorian, 1999] Gregorian, R. (1999). *"Introduction to CMOS OP-Amps and Comparators"*. John Wiley & Sons, Inc.

[Haartsen and Mattisson, 2000] Haartsen, J. C. and Mattisson, S. (2000). "Bluetooth–A New Low-Power Radio Interface Providing Short-Range Connectivity". *Proceedings of IEEE*, 88(10):1651–1661.

[Hamade, 1978] Hamade, A. R. (1978). "A Single-Chip All-MOS 8-bit A/D Converter". *IEEE Journal of Solid-State Circuits*, 13(12):785–791.

[Hendriks, 1997] Hendriks, P. (1997). "Specifying Communication DAC's". *IEEE Spectrum*, 34(7):58–69.

[Hogervorst et al., 1994] Hogervorst, R., Tero, J. P., Eschauzier, R. G. H., and Huijsing, J. H. (1994). "A Compact Power-Efficient 3 V CMOS Rail-to-Rail Input/Output Operational Amplifier for VLSI Cell Libraries". *IEEE Journal of Solid-State Circuits*, 29:1505–1513.

[Holloway, 1984] Holloway, P. (1984). "A Trimless 16-bit Digital Potentiometer". *IEEE Intl. Solid-States Circuits Conf.*, pages 66–67.

[Hotta and *et al.*, 1987] Hotta, M. and *et al.* (1987). "A 12-mW 6b Video Frequency ADC". *IEEE Int. Solid-State Citcuits Conf.*, pages 100–101.

[Jones and Martin, 1996] Jones, D. and Martin, K. (1996). *"Analog Integrated Circuit Design"*. John Wiley and Sons, Inc.

[Kim et al., 1997] Kim, K. Y., Kusayanagi, N., and Abidi, A. (1997). "A 10-b, 100-MS/s CMOS A/D Converter". *IEEE Journal of Solid-State Circuits*, 32(3):302–311.

[Kuboki and *et al.*, 1982] Kuboki, S. and *et al.* (1982). "Nonlinearity Analysis of Resistor String A/D Converter". *IEEE Trans. Circuits and Systems*, CAS-29:383–389.

[Lewis and Gray, 1987] Lewis, S. H. and Gray, P. R. (1987). "A Pipelined 5-Msample/s 9-bit Analog-to-Digital Converter". *IEEE Journal of Solid-State Circuits*, 22(12):954–961.

[Li, 2001] Li, X. (2001). *"System and Circuit Design Techniques for CMOS Soft-ware Radio Receiver Front-end"*. Ph.D. Dissertation, The Ohio State University.

[Lin and Bult, 1998] Lin, C.-H. and Bult, K. (1998). "A 10-b, 500-MSample/s CMOS DAC in 0.6 mm^2". *IEEE Journal of Solid-State Circuits*, 33(12):1948–1958.

[Lynn and Ferguson, 1997] Lynn, L. and Ferguson, P. J. (1997). "A Capacitor-Based D/A Converter with Continuous Time Output for Low-Power Applications". *International Symposium on Low-Power Electronic Designs*, pages 119–124.

[Matsuya and Tamada, 1994] Matsuya, Y. and Tamada, J. (1994). "1V Power Supply Low-power Consumption A/D Conversion Techniques with Swing Suppression Noise Shaping". *IEEE Journal of Solid-State Circuits*, 29(12):1524–1530.

[Mehr and Singer, 1999] Mehr, I. and Singer, L. (1999). "A 55-mW, 10-bit, 40 Msample/s Nyquist Rate CMOS ADC". *Proc. IEEE Custom Integrated Circuits Conf.*, pages 113–116.

[Miki et al., 1986] Miki, T., Nakamura, Y., Nakaya, M., Asai, S., Akasaka, Y., and Horiba, Y. (1986). "An 80-MHz 8-bit CMOS D/A Converter". *IEEE Journal of Solid-State Circuits*, sc-21(6):983–988.

[Nakamura et al., 1991] Nakamura, Y., Miki, T., Maeda, A., Kondoh, H., and Yazawa, N. (1991). "A 10-b 70-MS/s CMOS D/A Converter". *IEEE Journal of Solid-State Circuits*, 26(4):637–642.

[Oborn, 1996] Oborn, P. K. (August 1996). *"Digital to Analog Converter Design Using Modified Resistor String Architecture"*. Thesis, Brigham Young University.

[Peetz et al., 1986] Peetz, B., Hamilton, B. D., and Kang, J. (1986). "An 8-bit 250Megasample Per Second Analog-to-Digital Converter: Operation without a Sample and Hold". *IEEE Journal of Solid-State Circuits*, 21(12):997–1002.

[Pelgrom, 1990] Pelgrom, M. J. M. (1990). "A 10-b 50-MHz CMOS D/A Converter with 75-Ω Buffer". *IEEE Journal of Solid-State Circuits*, 25(12):1347–1352.

[Pelgrom et al., 1989] Pelgrom, M. J. M., Dunimaijer, A. C. J., and Welbers, A. P. G. (1989). "Matching Properities of MOS Transistors". *IEEE Journal of Solid-State Circuits*, 24:1433–1439.

[Peluso et al., 1998] Peluso, V., Vancorenland, P., Marques, A. M., Steyaer, M. S. J., and Sansen, W. (1998). "A 900-mV Low-power $\Delta \sum$ A/D Converter with 77-dB Dynamic Range". *IEEE Journal of Solid-State Circuits*, 33(12):1887–1897.

[Razavi, 1995] Razavi, B. (1995). *"Principles of Data Conversion System Design"*. IEEE Press.

[Rudell et al., 1997] Rudell, J. C., Qu, J. J., Cho, T. B., Chien, G., Brianti, F., Weldon, J. A., and Gray, P. R. (1997). "A 1.9-GHz Wide-Band IF Double Conversion CMOS Receiver for Cordless Telephone Applications". *IEEE Journal of Solid-State Circuits*, 32(12):2071–2088.

[Sansen et al., 1998] Sansen, W., Steyaert, M., Peluso, V., and Peters, E. (1998). "Toward sub-1-V Analog Integrated Circuits in Submicron Standard CMOS Technologies". *Proc. IEEE Int. Solid-State Cicruit Conf.*, pages 186–187.

[Schouwenaars et al., 1988] Schouwenaars, H. J., Groeneveld, D. W. J., and Termeer, H. A. H. (1988). "A Low-Power Stereo 16-bit CMOS D/A Converter for Digital Audio". *IEEE Journal of Solid-State Circuits*, 23(6):1290–1297.

[Shi et al., 2001a] Shi, C., Wilson, J., and Ismail, M. (2001a). "Design Techniques for Improving Intrinsic Accuracy of Resistor String DACs". *IEEE International Symposium on Circuits and Systems*.

[Shi et al., 2001b] Shi, C., Wu, Y., Lin, C.-H., and Ismail, M. (2001b). "Design and Power Optimization of High-Speed Pipeline ADC for Wideband CDMA Applications". *Journal of Integrated Circuits and Signal Processing*, 26(3):229–238.

[Shu et al., 1996] Shu, T. H., Bacrania, K., and Gokhale, R. (1996). "A 10-b 40-Msample/s BiCMOS A/D Converter". *IEEE Journal of Solid-State Circuits*, 31(10):1507–1510.

[Tadjpour et al., 2001] Tadjpour, S., Cijvat, E., Hegazi, E., and Abidi, A. (2001). "A 900MHz Dual Conversion Low-IF GSM Receiver in 0.35μm CMOS". *ISSCC 2001*, pages 292–293.

[Van den Bosch et al., 2000] Van den Bosch, A., Steyaert, M., and Sansen, W. (2000). "An Accurate Statistical Yield Model for CMOS Current-Steering D/A Converters". *IEEE International Symposium on Circuits and Systems*, pages IV–105–IV–108.

[Van der Plas et al., 1999] Van der Plas, G. A. M., Vandenbussche, J., Sansen, W., Steyaert, M. S. J., and Gielen, G. G. E. (1999). "A 14-bit Intrinsic Accuracy Q^2 Random Walk CMOS DAC". *IEEE Journal of Solid-State Circuits*, 34(12):1708–1718.

[Vorenkamp and Roovers, 1993] Vorenkamp, P. and Roovers, R. (1993). "An 12-b, 60-Msample/s Cascoded Folding and Interpolating ADC". *IEEE Journal of Solid-State Circuits*, 28(3):292–300.

[Waltari and Halonen, 2001] Waltari, M. and Halonen, K. A. I. (2001). "1-V 9-Bit Pipelined Switched-Opamp ADC". *IEEE Journal of Solid-State Circuits*, 36(1):129–134.

[Wu et al., 1996] Wu, J. T., Chang, Y. H., and Chang, K. L. (1996). "A 1.2 V CMOS Switched-capacitor Circuits". *IEEE International Solid State Circuits Conference*, pages 388–389.

[Yoshii and *et al.*, 1987] Yoshii, Y. and *et al.* (1987). "An 8b 350MHz Flash ADC". *IEEE Int. Solid-State Circuits Conf.*, pages 96–97.

Index

A

AC performances, 16
Active series switch, 63
Amplifier settling errors, 24
Analog baseband chain, 3
Analog-to-digital converters (ADCs), 2

B

Bluetooth, 1
Bottom-plate sampling, 41

C

Calibration, 101
Capacitor mismatch, 73
Capacitor scaling, 34
Cascode compensation, 75
Class-AB output stage, 109
Clock boosting, 59
Clock generator, 47
Common-mode feedback (CMFB), 45
Control DACs, 103
Current steering DACs, 91

D

Data converters, 2
DC offset, 10
Deglitching circuit, 111
Differential nonlinearity (DNL), 15
Digital correction circuit, 84
Digital error correction, 25
Digital-to-analog converters (DACs), 2
Direct conversion, 9
Dynamic averaging, 101
Dynamic biasing, 35
Dynamic comparator, 33
Dynamic range, 54

E

Effective number of bit (ENOB), 73
ETSI (European Telecommunications Standards Institute), 39

F

Feedback factor, 85
FH/time division duplex (FH/TDD), 72
Flash ADC, 17
Folded resistor string, 104
Folded-cascode, 43
Form factor, 1
Frequency hopping (FH) spread spectrum, 72
Frequency synthesizer, 8

G

Gain error, 15
Gain-bandwidth product (GBW), 30
Gaussian-shaped frequency shift keying (GFSK), 72
Glitch energy, 111
Global System for Mobile communication (GSM), 4
Gradient errors, 95

H

High-resolution, 4
High-speed, 4
HomeRF, 71
Homodyne, 9

I

IEEE 802.11 standard, 39
Image rejection (IR) filter, 8
IMT-2000 (International Mobile Telecommunication 2000), 103
Industry-Scientific-Medical (ISM), 39
Integral nonlinearity (INL), 15
Interpolating and folding ADCs, 19

L

Least significant bits (LSBs), 92
Level shifting, 65
LO-to-antenna leakage, 10
Low noise amplifier (LNA), 7
Low voltage operation, 5
Low-V_T devices, 58
Low-IF, 9

M

Major code transition, 111
MDAC (multiplying DAC), 22
Mixer, 7
Monolithic integration, 7
Most significant bits (MSBs), 92
Multi-standard, 4
Multiple resistor string architecture, 105

O

Offset errors, 24
Out-of-band energy, 8
Oversampling ADCs, 19

P

Pipeline ADC, 20
Power consumption, 1
Power optimization techniques, 32
Power optimization, 5
Power-efficient, 5
Programmable capacitor array (PCA), 93

R

Rail-to-rail operation, 57
Random error, 95
Resistor string DACs, 94
Resolution, 15

S

Sample-and-hold, 20
Selectivity, 9
Semiconductor Industry of Association (SIA), 2
Sensitivity, 9
SFDR (Spurious Free Dynamic Range), 16
Sigma-delta modulator, 19
Slew rate, 31
Smart-biasing, 49
SNDR (Signal to Noise plus Distortion Ratio), 17
SNR (Signal to Noise Ratio), 16
Special interest group (SIG), 72
Static (DC) performance, 15
Sub-ADC, 22
Switched-capacitor DAC, 93
Switched-capacitor, 40
Switched-opamp, 60
System-on-chip (SoC), 5

T

Telescopic OTA, 43
THD (Total Harmonic Distortion), 17
Thermal noise, 24
Three-phase non-overlapping clock, 46
Trimming, 101
Two-phase non-overlapping clock, 41
Two-step ADC, 19

U

UMTS (Universal Mobile Telecommunication System), 103

V

VGA, 12

W

WCDMA (Wideband Code Division Multiple Access), 39
Wideband IF double conversion, 9
Wireless communication, 1
Wireless local area network (WLAN), 1

Z

Zero-IF, 9